Reliability Investigation of LED Devices

for Public Light Applications

**Durability, Robustness and Reliability of
Photonic Devices Set**

coordinated by
Yannick Deshayes

Reliability Investigation of LED Devices for Public Light Applications

Raphael Baillot
Yannick Deshayes

ELSEVIER

First published 2017 in Great Britain and the United States by ISTE Press Ltd and Elsevier Ltd

ISTE Press Ltd
27-37 St George's Road
London SW19 4EU
UK

Elsevier Ltd
The Boulevard, Langford Lane
Kidlington, Oxford, OX5 1GB
UK

www.iste.co.uk

www.elsevier.com

Notices
Knowledge and best practice in this field are constantly changing. As new research and experience broaden our understanding, changes in research methods, professional practices, or medical treatment may become necessary.

Practitioners and researchers must always rely on their own experience and knowledge in evaluating and using any information, methods, compounds, or experiments described herein. In using such information or methods they should be mindful of their own safety and the safety of others, including parties for whom they have a professional responsibility.

To the fullest extent of the law, neither the Publisher nor the authors, contributors, or editors, assume any liability for any injury and/or damage to persons or property as a matter of products liability, negligence or otherwise, or from any use or operation of any methods, products, instructions, or ideas contained in the material herein.

For information on all our publications visit our website at http://store.elsevier.com/

British Library Cataloguing-in-Publication Data
A CIP record for this book is available from the British Library
Library of Congress Cataloging in Publication Data
A catalog record for this book is available from the Library of Congress
ISBN 978-1-78548-149-9

Printed and bound in the UK and US

Contents

Introduction

The technological evolution of optoelectronic components such as light-emitting diodes (LEDs), particularly for applications related to lighting, introduces the challenge of increasing reliability and the optical power at a millimeter volume, while reducing manufacturing costs. The standards for public lighting today impose a lifetime of over 50,000 h.

However, the different phases of development of these components must meet several criteria: minimizing the concentration of defects in the chip's materials, performing quality control of the interfaces between the various epitaxial layers, and ensuring surface quality for optimal light emission. Assembly processes must therefore be realized with high-performance materials in terms of heat dissipation and allow the design of a device that can both significantly increase the extraction of light and protect the emitting chip against external factors (temperature, vibration, chemical pollution, etc.).

In general, the major failure mode characterizing an LED is the gradual degradation of the optical power after aging. The demonstration of the failure mechanism at the origin of this degradation requires the implementation of physical analysis means that can be relatively long and costly. A complementary approach is to estimate the behavior in the form of an equivalent electro-optical model in a static regime. The parameters of this model are thus dependent on technology and their evolution reflects an electrical and/or optical decline.

The complexity of an LED-type optoelectronic component, relating to multiple methods used for manufacturing the chip and/or the various assembly phases, makes the identification of the element responsible for the optical power decline even more difficult.

The qualification standards currently required cannot be assured by the usual selection techniques of finished products, nor demonstrated by the accelerated tests which assess the average lifetime (MTTF). Indeed, even by accepting only two failures in a standard test of 1,000 h with an acceleration factor of 300–400, several hundreds or thousands of components are required in the composition of test samples. In public lighting, the JEITA or MIL-type qualification standards require a minimum number of components ranging from 30 to 100 depending on the type of aging applied. Regarding the actual failure, there are non-destructive analysis methods requiring little or no sample preparation: infrared thermography for a component's heat dissipation mapping, thermal imaging of a component by laser reflectometry, the current–voltage characteristic of the emissive component, the spectral analysis of the light emitted through the optical lens or the analysis of the output optical power.

The techniques for electro-optical analysis, regrouping the current–voltage characteristics I(V), the spectral characteristics L(E) and optical power P(I), are commonly used in the industry. However, the wealth of information contained in the latter is rarely used or controlled. In general, the main problem lies in the interpretation of results that often requires failure mechanisms or precise knowledge of the technology and architecture of the component and assembly. The measurement systems are then used systematically, allowing a verification of a "Go/No Go" type.

This work therefore aims to consider the optoelectronic component as an entity having a dual definition:

– a definition based on physical modeling from complex equations governing its functionality related to technology;

– a simpler to use "system"-oriented definition, based on simplified models with a set of more restricted parameters taking into account the interaction between the component and its environment.

Our study, favoring the second definition, is based on a failure analysis methodology capable of extracting the failure mechanism(s) responsible for the degradation of the component. This methodology is based on a growing demand from assembled LED manufacturers, which can meet certain needs in terms of characterization and aids in the reliability forecast by providing fault indicators for revealing physical degradation phenomena induced by environmental stresses. These needs therefore require a rapid and reliable methodology using a limited number of samples and information on the studied component. The latter are the parameters provided by the manufacturers' technical documentation or a set of easily measurable quantities. The objective of this methodology is threefold:

– to identify all information regarding the materials constituting the component and its assembly using the information given by the manufacturer and a set of physical and chemical analyses that can sometimes require a sample preparation. This phase will model the component from electro-optical and thermal viewpoints in order to extract the physical parameters that will address the second objective of this methodology;

– to pre-locate degraded areas by using electro-optical and thermal characterization for extracting electrical, optical and thermal failure signatures. These indicators are used to locate the faulty parts of the chip or its assembly;

– to confirm these damaged areas from physicochemical analyses for the suitable materials being characterized and for the level of information to be extracted.

This study aims to demonstrate that the methodology developed as part of this research can be applied to both a component's different design phases and various types of assembled components.

The book is structured into four chapters:

– the first chapter introduces the research of the state-of-the-art market for LEDs and different technologies based on GaN. We continue by describing the economic context and the technologies discussed in this book. All these elements will justify our study relative to national and international actors in the field;

– the second chapter recalls the physical principles involved in GaN technology by linking transport phenomena and electronic transitions to the component's functional parameters. The latter allow the establishment of physical models, equivalent to the component, from analyses performed on the complete system and taking into account the results from the literature. A reminder of the principle of each physicochemical analysis is also presented. We will further emphasize our presentation of the type of analyzed materials and the results of these analyses. A classification of physicochemical analyses will be proposed at the end of the chapter to reinforce the important link between a supposedly degraded area and the appropriate analysis means;

– the implementation of the methodology on low-power (<30 mW) GaN LEDs will be the main subject of the third chapter. Therefore, this chapter shows the adaptation of the methodology to build so-called "operational" reliability;

– finally, the fourth and final chapter presents a study conducted in collaboration with a national assembler, for the evaluation of the failure mechanism involved in the yellowing of the white light power LEDs used in public lighting. The aim of this chapter is to demonstrate that the methodology can be integrated from the component's design (so-called "constructed" reliability). This theme comes in agreement with the methodology for the construction of reliability defined by the EDMINA (Evaluation of Micro and Nano Assembled Devices) research team with which this research has been carried out.

1

Light-emitting Diodes: State-of-the-Art Gan Technologies

The incredible growth of GaN technologies for the manufacture of optoelectronic components in many fields of application (medical, energy and information and communication techniques) has led to a booming LED market. Since the 2000s, the concern of our society about the environment and particularly the reduction of power consumption have led to a very strong growth in the field of public lighting. The GaN-based LED has entered the market previously controlled by the giants of fluorescent lighting and incandescent lamps. This leads to a replacement of these lamps by LED lamps. The miniaturization of LED lighting devices and the increase in their performance (> 150 lm.W^{-1}) has led to an increase in power density. Thus causing a resurgence of challenges related to thermal flow and lifetime.

The increasing development of the complexity of technologies based on Gallium Nitride (GaN) and the miniaturization of assembly technologies make failure analysis difficult. The direct consequence is that the reliability of these systems is more difficult to estimate. Many manufacturers base their estimates on an exponential mathematical projection for evaluating lifetimes of more than 50,000 h. However, most of them base their evaluations on the feedback of incandescent lamps [SMI 10]. Currently, LED lamps are still of too low luminance and their color slightly shifts towards blue. This difference with conventional lamps leads to a market equilibrium. For LED lamp manufacturers, the study of reliability, specifically the analysis of physical failure, is very critical and is becoming a selling point of equal importance to the reduction of energy consumption.

This chapter will provide the economic context in which the research lies by positioning the issues and challenges of the market for GaN LED technology. After this first part, the objectives of the study will be defined. Thereafter, the state-of-the-art GaN technology will be developed. We will specify the physical properties of nitride materials, the structures associated with the components and the architecture of the assemblies for LEDs at low and high optical powers. Finally, the international and national positions of this research will be proposed along with its position within the framework of the objectives and concepts of the EDMINA research team, within which the studies presented in this book were performed.

1.1. Current economic context

Light-emitting diodes are electro-optical transducers allowing varied applications nowadays. The different areas addressed by LEDs are medical, energy, and information and communications technology (ICT). The technology currently used for LEDs cover the emission wavelength band ranging from ultraviolet (350 nm) to infrared (2,000 nm). This wavelength band makes it possible to address many societal needs.

The strong development of GaN technologies during the past 15 years has allowed the consideration of applications using the ultraviolet-visible band. The applications associated with the visible band will be specifically developed. The market associated with this relatively new technology is emerging. This section provides an overview of the market for GaN LEDs, especially of the most prevalent technologies in this market. This economic environment will help to frame the scientific objectives of the study.

1.1.1. Global LED market

The global market for LEDs has been experiencing a boom since the late 1990s. It is governed by a growing demand for increasingly reliable LEDs with an increase in production volumes for the sectors of television screens, LED lighting (TV backlight), laptops, mobile phones and lighting. On a global scale, focusing on the reduction in energy consumption envisions a bright future for LEDs, especially in the lighting of residential (private) and commercial (professional) buildings. In 2010, it exceeded the threshold of 10 billion US dollars (USD) [LED 10a, HSU 10].

With an overall annual growth of 13.6% from 2001 to 2012, this market will reach a record $ 14.8 billion USD by 2015 [GLO 10]. Figure 1.1 shows the evolution of the global LED market from 2001 to 2012 [ISU 07].

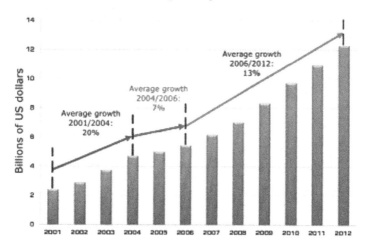

Figure 1.1. *Evolution of the global LED market from 2001 to 2012*

From 2001 until today, the market growth is estimated at 78% with an average increase of 13.9% per year. Three growth stages are observed:

– from 2001 to 2004, annual growth of 20%;

– from 2004 to 2006, corresponding to a period of unfavorable global economic context (annual growth lowered to 7%);

– a market recovery from 2006 onwards with an average annual growth of about 13%.

The major players in the LED industry are present in six economic powers: Europe, Japan, Korea, Taiwan, China and the United States. Figure 1.2 shows the distribution of the LED market revenues between these economies for the years 2009 and 2010 [LED 10b].

Japan is an important part of the global market for LEDs, as it contains two of the largest suppliers in the market: Nichia and Toyoda Gosei [GLO 10]. From 2007 until today, Nichia has always remained the leader among the 10 largest LED suppliers. Table 1.1 shows the top 10 LED global suppliers according to market analyst J. Hsu [HSU 10].

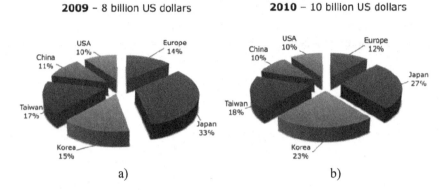

2009 – 8 billion US dollars   **2010 – 10 billion US dollars**

a)   b)

Figure 1.2. *Distribution of the LED market revenues*
of sixprincipal stakeholder countries: a) 2009 and b) 2010

Rank	2007		2008		2009		2010	
1	Nichia	24.0%	Nichia	19.0%	Nichia	16.0%	Nichia	15.0%
2	Osram	10.5%	Osram	11.0%	Osram	10.0%	Samsung	10.0%
3	Lumileds	6.5%	Lumileds	7.0%	Samsung	6.5%	Osram	9.0%
4	Seoul S.	5.0%	Seoul S.	5.5%	Lumileds	6.0%	Seoul S.	7.5%
5	Citizen	5.0%	Everlight	4.0%	Cree	5.5%	Cree	6.0%
6	Everlight	4.5%	Citizen	4.0%	Seoul S.	5.5%	Lumileds	5.5%
7	Stanley E.	3.5%	Cree	4.0%	Everlight	4.5%	Sharp	5.5%
8	Kingbright	3.5%	Stanley E.	3.0%	Stanley E.	4.5%	LG Inno.	4.5%
9	Avago	3.5%	Kingbright	3.0%	Lite-ON	3.5%	Everlight	4.0%
10	Toshiba	3.5%	Avago	3.0%	Citizen	3.0%	Stanley E.	3.5%
	Others	30.5%	Others	35.5%	Others	35%	Others	29%
	Total	100.0%	Total	100.0%	Total	100.0%	Total	100.0%

Table 1.1. *Top 10 of global LED suppliers from*
2007 to 2010 according to market share (%)

The companies Nichia and Toyoda Gosei (Japan), Philips Lumileds, Cree (USA) and Osram (Europe) are considered to be the five major players in the LED market; Toyoda Gosei is particularly present in the power GaN LED market. These stakeholders are mainly foundries and master all the manufacturing stages of a chip. For indoor/outdoor lighting applications, backlighting of TV screens, mobile phones, tablets or laptops, they can provide end users with "ready to use" components. We also recognize the strong presence of assemblers from four of the six economic powers involved in this market: Sharp, Toshiba, Citizen and Stanley for Japan, Avago for the United States, Lite-On, Everlight and Kingbright for Taiwan, and LG, Samsung LED and Seoul Semiconductor for Korea.

Such growth has led to the increase in wafers' size for GaN chip production: 53% in 2010 and 71% in 2011. This led to a boom in the market for the production of chips by organometallic vapor phase epitaxy (MOCVD). As a result, 25 new businesses were created between 2010 and 2011 [LED 10a]. There are now over 75 LED manufacturers in the world. China, a leading manufacturer of MOCVD since 2010, represents today one of the largest production areas. It particularly encourages Korea and Taiwan to implant their production areas in Chinese territories [HSU 10].

1.1.2. Societal and market issues of GaN LEDs for public lighting

The increase in population on a global scale (> 9 billion by 2050 against 7 billion in 2011) and the increasingly demanding preservation of natural heritages lead today to a clear need for a reduction in energy consumption [CEN 09]. To meet this need, new technological solutions are emerging in many areas, particularly in the field of LEDs. These components have undergone, since the early 1990s, considerable growth mainly due to the emergence of new markets such as that of public lighting. This success is explained by the great variety of sectors in which LEDs play a role. Public lighting, mobile phones, signage, TV screens (backlight), automotive, medicine and military (niche areas) account for the majority of LED market sectors. Figure 1.3 shows the market distribution in 2010 by sector [HSU 10].

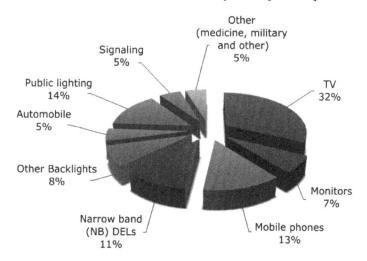

Figure 1.3. *LED market distribution as a function of sectors for the year 2010*

In all of these applications, GaN-based LED technologies are dominant and lead the market. This is explained by the predominance of these technologies in TVs, public lighting, automobiles, mobile phones, monitors and signage. Figure 1.4 shows the market evolution of LEDs according to their technologies (GaAs/GaP, AlInGaP and GaN) from 2006 to 2010 [HSU 10].

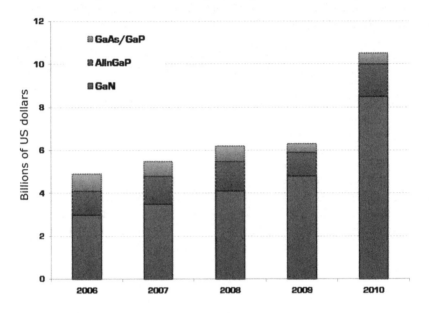

Figure 1.4. *Evolution of the LED market as a function of different technologies (GaAs/GaP, AlInGaP and GaN) from 2006 to 2010*

J. Hsu *et al.* noted that the production of GaN LEDs has increased by 76% and that this technology represents 81% of the global LED market today [HSU 10].

This tremendous growth is the result of both economic and environmental challenges that are driving the lighting market. Table 1.2 provides the projections of the Optoelectronics Industry Development Association (OIDA), which specifies the key issues in the field of LED lighting [OID 10].

The major advances in GaN-based LED technologies are based on four major issues: the optical power, energy efficiency, quality of light (CRI) and longer lifetimes (> 50,000 h).

Technology	LED (2002)	LED (2007)	LED (2012)	LED (2020)	Incandescent lamp	Fluorescent lamp
Light output (lm.W^{-1})	25	75	150	200	16	85
Life cycle (kh)	20	>20	>100	>100	1	10
Flux (lm/lamp)	25	200	1,000	1,500	1,200	3,400
Optical power (W/lampe)	1	2.7	6.7	7.5	75	40
Cost of lumens ($/klm)	200	20	<5	<2	0.4	1.5
Cost of a lamp ($/lamp)	5	4	<5	<3	0.5	5
Color output index (CRI)	75	80	< 80	<80	95	75
Penetratedlightingmarkets	Weakflux	Incand	Fluor	All	–	–

Table 1.2. *Overview and projections for the lighting and power LED market (OIDA)*

In general, an LED lamp of 13 W emits as much light as an incandescent lamp of 100 W. In Japan, lighting accounts for 16% of electricity consumption. According to the Institute of Energy Economics of Japan (IEEJ), if all Japanese lamps were replaced by LED lighting, electricity consumption in Japan would be reduced by 9%. The success of LEDs in public lighting lies in the fact that they represent a "green" alternative meeting certain environmental criteria (reduction in energy consumption, mercury and lead-free technologies). Despite the existence of ECO fluorescent tubes that can save 50% of energy compared with a conventional tube, Frost & Sullivan identifies quality assurance at a reasonable price as a key issue for the LED lighting market [FRO 11a]. The OIDA guidelines predict the cost of an LED lamp to be below 3 USD in 2020 for an optical power greater than 7 W.

GaN power LEDs are therefore now experiencing a boom associated with that of the public lighting market. Figure 1.5 shows the penetration of the GaN power LED market in the global public lighting market [FRO 11b].

The market research firm Frost & Sullivan reported today (2011) that the market for GaN LEDs has penetrated the public lighting sector to about 3% (Figure 1.5). Legislation and regulation in public lighting represents a critical factor for the evolution of this market in the next decade. Indeed, the public lighting market is a four-speed market that is localized and segmented into four main regions: Europe, the North American Free Trade Agreement (NAFTA) composed of the United States, Canada and Mexico, Asia-Pacific (APAC) which groups the Far East, the Indian subcontinent and Oceania, and the rest of the world (ROW). Figure 1.6 shows the legislative directives of these four groups for the replacement of traditional lighting by LED lighting [RES 11].

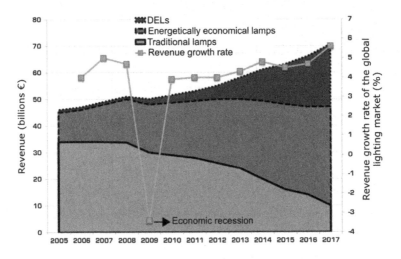

Figure 1.5. *Evolution of the global public lighting revenues and their growth rates from 2005 to 2017 by type of lamps*

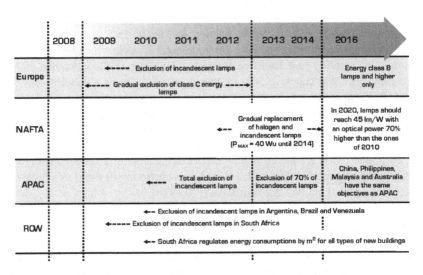

Figure 1.6. *Legislative roadmap from 2008 to 2020 for the replacement of tradition lighting by LED lighting*

Despite the recent economic downturn in 2009 (Figure 1.5) involving a slowdown in consumer consumption, the LED lighting market is set to have significant growth in the next decade. Figure 1.7 forecasts this growth in terms of revenue until 2017 [FRO 11a].

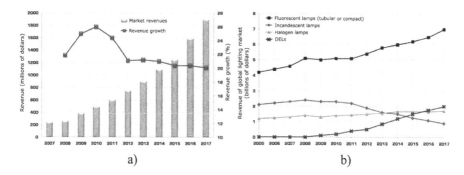

a)   b)

Figure 1.7. *a) Evolution of the GaN power LED market (revenues and growth rate) for public lighting from 2007 to 2017. b) Revenues of the global lighting market by types of lamps*

The dominance of fluorescent lamps in lighting is evident. This is explained by their cost being significantly lower than that of LED lamps ($ 0.5 per tube against approximately $ 5 for an LED lamp). However, in the medium term, the requirements in terms of energy efficiency of lamps are very promising for LED lamps, particularly owing to the constant improvement of their reliability (1,000 hours for a fluorescent tube against 25,000–50,000 h for LED lamps) and their functionality (wide range of applications). Frost & Sullivan estimates that, by 2017, the market for LED public lighting will have penetrated the global public lighting market to 31% (Figure 1.7(b)) with an annual growth rate of 21.3% from 2010 to 2017 (Figure 1.7(a)) [FRO 11a].

LED public lighting market stakeholders are grouped into three tiers:

– Philips, Nichia and Cree together provide half of the world production;

– other companies like Osram, Havells-Sylvania and GE Lighting are part of the second production tier;

– the remaining tier consists mainly of regional companies such as Megaman and Zumtobel (Europe), Illumisys (USA) and other small companies, mostly in Asia.

Table 1.3 shows the ranking of the top five stakeholders in the global market for public lighting in 2010 by region and type of market (lamps and energy saving) [FRO 11b].

Region	Type of market	Market size (billion €)	Rank				
			1	2	3	4	5
Europe	Lames	3.00	Philips	Osram	GE	Megaman	Panasonic
	ECO (control gear)	0.60	Panasonic	Osram	Philips	Zumtobel	Helvar
NAFTA	Lamps	2.80	GE	Osram	Philips	Acuity brands	Panasonic
	ECO (control gear)	1.00	Philips	UniversalLighting	Osram	GE	Acuity brands
APAC	Lamps	3.40	Philips	Osram	GE	Panasonic	NVC
	ECO (control gear)	1.00	Philips	Osram	Panasonic	Toshiba	GE
ROW	Lamps + ECO (control gear)	2.0	Philips	Osram	GE	Panasonic	Zumtobel

Table 1.3. *Top 5 of the public lighting market stakeholders by region and type of market in 2010*

We see that the leading suppliers of GaN power LEDs for public lighting are Philips, GE, Osram, Megaman, Panasonic and Zumtobel. The leader for the demand in this market is now Europe because European environmental guidelines encourage the reduction in energy consumption by investing in the replacement of incandescent and fluorescent lamps with LED lamps. For the same reason, another market report identifies market opportunities for LED lighting in Asia. It could reach 2.1 billion by 2016 with the Chinese market alone estimated to reach 420 million USD [RES 11]. Meanwhile, Japan is seen as the leading supplier in this market. By 2016, its sales will exceed one billion USD. Indeed, it responds to 41.3% of LED applications' demand in the Asian market and many large companies establish their headquarters in Japan. In Korea, sales should reach around $ 230 million by 2016 [NG 11]. Samsung and LG are considered to be the Korean leaders in the field of LEDs for public lighting.

Our society's concern about the environment, and particularly the reduction in energy consumption, has led to the emergence of new markets for GaN power LEDs for public lighting in Asia and in the rest of the world (ROW).

1.2. State-of-the-art GaN-based LEDs

GaN-based technologies have, in part, contributed to solving a relatively significant social challenge: saving electric power consumed by public lighting lamps. Since the early 2000s, these technologies have grown considerably. Considering all of these factors, the evolution of the GaN industry has been important over the past decade. Many technological solutions have been provided in order to improve both the efficiency ($lm.W^{-1}$) and the quality of the LED lamp's color rendering index (CRI). These developments in LED design occurred at both the chip and the assembly levels. Regarding the chip, it is mainly the optical efficiency that smelters have improved. The assembly of the LED, for its part, and, more broadly, of the lighting device must meet two criteria: the quality of the IRC and heat dissipation.

The state-of-the-art, described in this section, allows us to locate developments in the design of GaN LED devices. The key points needed to remove major technological barriers and thus meet the different needs associated with the societal issue described above will be proposed.

1.2.1. Nitrides: from the wurtzite structure to band engineering

GaN, as well as the associated binary compounds InN and AlN, is behind the development of many electronic and optoelectronic technologies. In electronics, the saturation velocity of GaN electrons is relatively large ($2.5.10^7$ cm.s^{-1}) and promotes the increase in the transition frequency of AlGaN/GaN electronic high-mobility transistors (HEMT) to up to 300 GHz. Other properties of GaN, such as high breakdown field ($> 5.10^6$ V.cm^{-1}), associated with a low rate of ionization by impact makes it possible to obtain a high output power (10 W.mm^{-1}at 40 GHz) [TAR 08]. The direct bandgap structure is also utilized to develop optoelectronic devices, allowing an emission spectrum that extends from the ultraviolet (UV) to the visible band. GaN is the only material that can have optoelectronic applications at short wavelengths (<500 nm). Many optoelectronic components, such as LEDs, laser diodes or detectors in hostile environments, metal–semiconductor–metal (MSM), are already available, and are a major research focus.

The difference in the radius between the gallium and nitrogen atom leads to the organization of the crystal in a wurtzite structure. This non-centrosymmetric structure leads to the existence of an internal electric field (> 5 MV.cm^{-1}) [COR 06]. This particular property has been used in all fields of applications. We can, for example, increase the speed of electrons in the structure in order to increase the transition frequency for electronic applications (HEMT).

In this part, the main structures used to develop LEDs will be described. The physical properties of wurtzite, GaN, AlGaN and InGaN ternary compounds will be studied. The growth of GaN on sapphire substrates and the electro-optical properties and engineering of bands of InGaN/GaN quantum well structures will be developed.

1.2.1.1. The wurtzite structure and physical properties of nitride materials

Three types of crystal structures exist in nitride materials: the wurtzite structure (Wz), zinc blende (ZB) and Rock Salt (HR). Under ambient conditions, the most thermodynamically stable structure is the Wz structure for AlN, GaN and InNbinary compounds [MOR 08a]. In contrast, the ZB structure is metastable and is mainly used for typologically compatible materials such as GaN and InN. Numerous studies have shown that this structure is stabilized by heteroepitaxial growth in thin films on crystal orientation planes (011) of cubic substrates, such as Si [LEI 91], SiC [PAI 89], MgO [POW 93] and GaAs [MIZ 86]. The RS structure, being rarer, may only be formed under a very high atmospheric pressure ranging from 12 to 52 GPa (120–520 kbar) according to the selected materials (InN, AlN or GaN) [XIA 93, PER 92, UEN 94]. This technique is too expensive and is only used in laboratories and not by the industry. Table 1.4 shows the values of the main electrical and optical parameters for GaN, AlGaN and InGaN compounds, the most commonly used in LED-type components [MOR 08a, PIP 02, PAN 09].

Parameters at 300 K	WzGaN	$Wz\,Al_x\,Ga_{1-x}\,N$	$Wz\,In_x\,Ga_{1-x}\,N$
E_g (eV)	3.42	4 (x = 0.3 and b = 0.7 eV)	2.17 (x = 0.3 and b = 3.8 eV)
μ_e $(cm^2.V^{-1}.s^{-1})$	< 1,000	35	300 – 850
μ_h $(cm^2.V^{-1}.s^{-1})$	< 200	9	< 15
N_c (cm^{-3})	$2.3.10^{18}$	5.10^{18}	$10^{16} – 10^{17}$
N_v (cm^{-3})	$4.6.10^{19}$	$1.5.10^{19}$	5.10^{18}
D_n $(cm^2.s^{-1})$	25	–	–
D_p $(cm^2.s^{-1})$	5–94	–	–
ε	9.5	–	10.9
n	~ 2.85	–	2.6 – 3

Table 1.4. *Electrical and optical parameters of GaN, AlGaN and InGaNwurtzite structure materials*

E_g, material gap; μ_e, μ_h, electron and hole mobility, respectively; N_C, N_V, effective state density in the conduction band and the valence band, respectively; D_N, D_P, diffusion coefficient of electrons and holes, respectively; ε, dielectric constant; n, optical index and b, band curvature parameter. The mobility values, concentrations as well as the gap depend on the x composition of each component. Table 1.5 gives the dependence on x of the lattice parameters (a and c) and of the gap for the AlGaN and InGaN materials [ZOR 01].

Material	Latticeparameters	Gap energy
$Wz\,Al_x Ga_{1-x} N$	$a_{Al_x Ga_{1-x} N} = 3.1986 - 0.0891x$	$E_g(x) = x.E_g(AlN) + (1-x).E_g$
	$c_{Al_x Ga_{1-x} N} = 5.2262 - 0.2323x$	$(GaN) - b_{AlGaN} x(1-x)$
$Wz\,In_x Ga_{1-x} N$	$a_{In_x Ga_{1-x} N} = 3.1986 - 0.3862x$	$E_g(x) = x.E_g(InN) + (1-x).E_g$
	$c_{In_x Ga_{1-x} N} = 5.2262 - 0.574x$	$(GaN) - b_{InGaN} x(1-x)$

Table 1.5. *Dependence on x of the lattice and gap parameters of AlGaN and InGaN materials*

The ternary compounds formed from WzGaN, InN and AlN materials enable a wide range of energy gaps with a small variation in the lattice parameters. Thus, control of the x composition of these compounds can build heterostructures, quantum wells or superlattices, giving rise to interesting electro-optical properties for LED-type components and laser diodes.

1.2.1.2. Band engineering and base structures

Band engineering is a technique used for the structuring of electronic and optoelectronic components. Usually, the horizontal axis is represented by the x dimension and the vertical axis by the energy of the carriers. In this case, we look at the gap of the material and at the variation in valence band energy levels and conduction as a function of x. This representation is very convenient for understanding electronic transport within a structure. In the case of optoelectronic structures, materials associated with epitaxy are of a different nature. This manufacturing process is called hetero-epitaxy and represents the main development technique of optoelectronic components.

In order to understand the optical phenomena within a single structure, the band diagram in real space is no longer usable. In general, only the material of the active region, light emitting area, is the result of electronic transitions. We then use the band diagram as a function of the wave vector k on a period of the crystal lattice: the first Brillouin zone. This representation, from a chemistry point of view, allows the highlighting of the parabolic bands and, thus, the shape of the spectral emission of an LED.

GaN is a material that is very often used in the active areas of LEDs. The wide gap of 3.42 eV provides a luminescence in the blue/UV range. Figure 1.8 shows the band diagram of the structure on the first Brillouin zone of WzGaN [IOF 01].

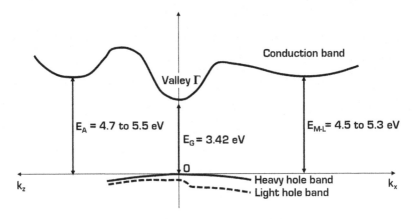

Figure 1.8. *Band diagram of the solid GaN material in Wz structure in the first Brillouin zone*

Many materials such as GaAs, InP, GaP or SiC have their conduction and parabolic valence bands in the Γ valley. The specificity of GaN is that the valence bands for heavy and light holes are not parabolic [IOF 01]. Chapter 2 will highlight how the electrical and optical models will be impacted by this type of structure.

Most material growth techniques III–V in solid form, substrate and their crystal orientation, have been the object of many efforts to grow thin film nitride materials. Indeed, techniques of vapor phase epitaxy by hydrides (HVPE), and organometallic (OMVPE) and molecular beam (MBE) methods have greatly improved the quality of the layers of nitride materials. However, all of these epitaxy techniques are faced with two major problems: the lack of GaN substrates in nature and the difficulty of incorporating nitrogen and ammonia flow (NH3) of the level (>10 L.min^{-1}) required for the preparation of semiconductors based on nitride materials containing indium atoms such as InGaN [MOR 08a].

A major drawback of GaN is that it is not available naturally in large quantities. This is partly due to the low solubility (1%) of nitrogen in a solid gallium crystal and the high pressure of nitrogen vapor (1.6 GPa) in growth conditions (1,500°C in the vapor phase) [DEN 06]. The best alternatives to address these problems lie in the development of sapphire substrates, SiC or Si with lattice parameters compatible with those of GaN.

Component miniaturization has led to increasingly complex structures. When the size of the semiconductor approaches the de Broglie wavelength, electrical and optical properties are modified and materials become more susceptible to external conditions (temperature, stresses at the interfaces). Figure 1.9 shows a diagram of the systems used in optoelectronic components today to reduce the size and change the electro-optical properties [MOR 08a].

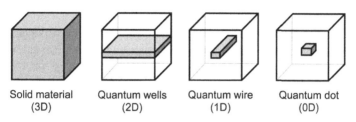

| Solid material | Quantum wells | Quantum wire | Quantum dot |
| (3D) | (2D) | (1D) | (0D) |

Figure 1.9. *Schematic representation of systems with different dimensions in real space*

When the thickness of a well in a structure is comparable to the de Broglie's wavelength, the state densities in the valence and conduction bands are said to be discrete. A quantum well is then formed whose physical properties are modified on the axis perpendicular to the growth axis (Figure 1.9). Figure 1.10 shows the band diagram of a multi-quantum well structure AlGaN/GaN.

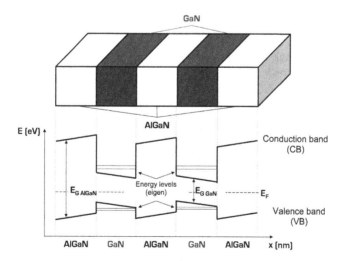

Figure 1.10. *AlGaN/GaN real multi-quantum well structure and its band diagram in real space*

This structure is composed of two AlGaN/GaN quantum wells: the AlGaN material is a region called bandgap and the GaN material created the potential well. A potential well is formed by the epitaxy of two semiconductor materials having a different gap. Electron transport is done primarily by electrons and holes. The quantum wells, in turn, promote electronic transitions, and thus, transport is no longer a case of diffusion as in a conventional structure.

The association of the two structures allows the optimization of the confinement of particles (electrons, photons) to the potential and the increase in the gain with the quantum wells. Electronic transitions occur spontaneously in a quantum well. When the structure includes several quantum wells (Figure 1.10), a phenomenon of amplification of light by stimulated emission is implemented and makes on increase in the internal quantum efficiency possible [ROS 02]. However, the very small width of a quantum well (about 10 Å) does not allow a good confinement of created photons and it is often necessary to combine a quantum well with a potential well [ROS 02]. The electron and the photon wave functions are almost zero outside the potential well. Therefore, the probability of presence of particles proportional to Ψ^2 is very low.

Several MPQ structures exist according to the selected wavelength. The PQs AlGaN/InGaN or GaN/AlGaN allow emissions in the UV (250–380 nm) while the PQs InGaN/GaN are used in the blue/green range (380–530 nm) [ROS 02, MOR 08c]. The latter are usually made by EJM or MOCVD on a GaN layer which is deposited on a substrate [SHA 03].

1.2.1.3. Usual substrates and doping materials for GaN

The native GaN substrate, or the one developed in the semiconductor industry, is very rare. It is therefore mandatory to use a host substrate for developing a component. The great difficulty is to find a substrate lattice parameter compatible with GaN. In the 1990s, the first technological barrier was lifted, even if the quality of the interface host substrate/GaN layer was not perfect (10^{10} defects per square centimeter) [MOR 08a]. To limit the spread of faults through the component, MOCVD-type buffer layers are produced by epitaxy. These techniques were mastered in the early 2000s, for the development of laser diodes that require an extremely low rate of defects ($<10^4$ cm^{-2}) [PAS 10]. The most commonly used materials as host substrates are SiC, Si and sapphire (Al$_2$O$_3$). Table 1.6 compares some physical properties of SiC, Si and Sapphire materials [MOR 08a, LEE 98, HIR 96, YUA 07].

Parameters at 300 K	SiC	Si	Sapphire
a (Å)	4.3596	5.43102	4.765
d (g.cm^{-3})	3.2	2.3290	3.98
T$_{fusion}$ (°C)	2,793	1,410	2,030
λ (W. cm^{-1}.K^{-1})	3.7	1.56	0.23
n	2.7	3.42	1.75
E$_G$ (eV)	2.36	1.12	8.1 – 8.6
T (%)	85 – 90	50 – 55	85 – 90
Transmission domain (μm)	0.4 – 0.8	1.5 – 6	0.2 – 5
ρ (Ω.cm)	$10^2 - 10^3$	$>50.10^3$	$> 10^{11}$

Table 1.6. *Physical parameters of SiC, Si and sapphire substrates*

d, density of the material; T$_{fusion,}$ melting temperature; λ, thermal conductivity; T, transmittance and ρ, electric resistivity.

The quality of solid SiC and its surface treatment, high thermal conductivity, its availability and high resistivity, explain the reason that SiC is a regularly used material for optoelectronic device substrates. Its weaknesses are the cost and the preparation it requires to yield an acceptable quality with a density of defects (dislocations) of less than 10^5 cm^{-2}[DMI 00]. Although some types of LEDs are manufactured on SiC substrates, the latter is mainly used to manufacture Field Effect Transistors (FET).

Silicon (Si) is the material which is particularly interesting in the manufacture of high-power LEDs. This is the least costly material, available in large wafers (300 mm), and, unlike GaAs, it has excellent thermal stability under the conditions established for GaN epitaxy. However, WzGaN formed on the Si substrate is very often highly defective (defect density of> 10^9 cm^{-2}) because the lattice mismatch is of the order of 17%, which leads to stresses in voltage [ZHO 05, LEE 05]. To overcome this problem, a very thin layer (<30 nm) of AlN, acting as a buffer layer, is generally used for growing the GaN transition layer [HON 02, TRA 99]. Indeed, the lattice mismatch between AlN and GaN is very low (2.5%) [NG 99].

Sapphire is considered to be the best compromise of the four substrates examined in Table 1.6. It is mainly used for low-power LEDs (<100 mW). In addition, sapphire is relatively inexpensive, transparent, available in large quantities for wafer sizes from 2 to 4 inches (or 6 inches) [RUB 11] and its quality in terms of surface defects in the solid material is continuously increasing. In addition, its transmittance remains high (> 85%) on the larger wavelength range (from 200 nm to 5 μm). This explains why it covers a wide range of applications in the manufacture of optoelectronic components (sensors, LEDs).

Concerning the doping in GaN layers, two key elements are widely used: silicon (Si) for an N-type doping and magnesium (Mg) for a P-type doping.

Si is a doping material that is mainly used to produce N-doped GaN layers. During the deposition phase by molecular beam, the physicochemical properties of silicon enable uniform and controlled doping in terms of concentration. The variation in the N doping concentration in silicon extends from 10^{17} to 2.10^{19} cm^{-3} by controlling the flow of SiH$_4$ by MOCVD [MOR 08a].

GaN P doping presents a second major technological obstacle in the development of optoelectronic devices based on GaN. Like any large gap material (> 3 eV), P doping has always been considered a difficult and delicate stage. Since 1989, the lack of P dopants for GaN-type nitrides has significantly slowed the progression of manufacturing optoelectronic devices. Two major obstacles have slowed the development of GaN technologies:

– the presence of hydrogen in the epitaxy techniques in the vapor phase. This creates a passivation layer on the Mg involving improper electrical and optical properties for the intended applications;

– the high P doping by Mg is difficult today because it is subject to self-compensation by the presence of defects created during the epitaxy acting as donors.

In the 1990s, it was shown that the presence of hydrogen is actually a great asset for obtaining a high electrical conductivity (P-type) [VAN 97]. However, the self-compensation phenomenon remains effective even today. The P area remains a critical area for this technology.

1.2.2. Electroluminescent GaN-based diodes

Investments in the GaN LED industry since the late 1990s caused considerable international growth of these components' technological evolution. This section thus provides a summary of technological improvements that have been made both in the "bare chip" and the assembly. Complex innovative techniques helped make GaN LEDs a technology able to respond today to needs that are both economic and environmental; primarily focused on reducing global electricity consumption.

1.2.2.1. *State-of-the-art of technological innovations of "bare chip" GaN LEDs*

GaN LEDs were created in the 1990s by three Japanese researchers (Isamu Akasaki and Hiroshi Amano, at Nagoya University in Japan, and Shuji Nakamura, who at the time worked at Nichia). These research efforts were awarded the Physics Nobel Prize in 2014. One of the major advances has been to master the thin film deposition of GaN on host substrates such as SiC and sapphire. A very large part of these research efforts focused on improving the epitaxial layers in wurtzite structures to reduce the defect rate in LED structures.

One of the first GaN LED structures is shown in Figure 1.11(a) [SHE 00]. The P contact usually consists of a very thin transparent bi-layer of Ni/Au (2 nm/6 nm) deposited on a doped GaN contact P layer of thickness 300 nm. The N contact is composed of a Ti/Al bi-layer (50 nm/2 µm) deposited on a doped GaN N layer. The latter is 3.5 µm thick and is deposited on a transition layer (30 nm) of un-doped GaN deposited on a sapphire substrate, electrical insulator, of thickness 80 µm. This high electrical conductivity property enables the development of a component with two superior contacts and reduces current leakages to the substrate. The active region comprises an MPQ $In_{0,3}Ga_{0,7}N/GaN$ (nine periods) structure manufactured by molecular beam epitaxy (MBE). The thickness of each InGaN well is 30 Å and that of each GaN barrier is 70 Å. The size of the chip deferral range is about 350 µm × 350 µm. This type of structure provides an optical power greater than 1.5 mW at 20 mA.

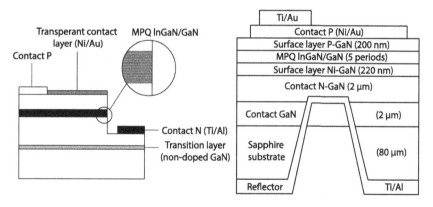

Figure 1.11. *Schematic representation of base structures of GaN LEDs*

Since the early 2000s, many efforts have been made to improve the optical power and the external luminescence efficiency and to reduce auto-heating. The design of vertical structure LEDs with a sapphire substrate engraved in "V" (Sapphire-Etched Vertical-Electrode Nitride Semiconductor – SEVENS) is presented in Figure 1.11(b) [KIM 05c, CHO 06].

With this technique, the external efficiency is improved by 8.4% compared with conventional structures, where the external side contacts yield is 7.5%. We can measure optical power of 4.5 mW at 20 mA [KIM 05c]. This improvement is attributed to the via etched into the sapphire substrate to reduce self-heating of the chip. The same structure was used for power LEDs whose optical power was 1.8–4.3 times greater than that of a conventional power LED powered at 200 mA [CHO 06].

Another work has helped to identify a third technological obstacle in improving the quality of sapphire substrates in order to increase optical power and external performance. Figure 1.12 summarizes all of the grounds for etching sapphire substrates [GAO 08, OH 08, WUU 06].

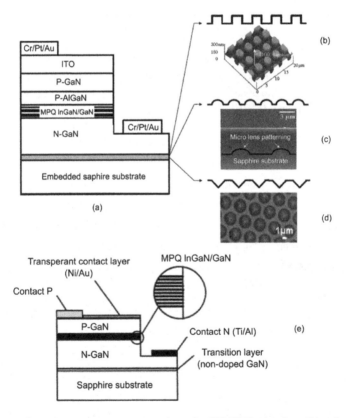

Figure 1.12. *a) Schematic representation of a SEVENS structure, b) cylindrical pattern and corresponding AFM image, c) spherical pattern with corresponding SEM image, d) pyramidal image with correponding SEM image, e) Schematic representation of conventionnal GaN LED*

We can obtain an increase in the optical power of between 17 and 37%, compared with a conventional structure (Figure 1.12(a)), using a cylindrical pattern (Figure 1.12(b)) engraved by wet and plasma ICP (inductively coupled plasma) [GAO 08]. Using the same pattern, we obtain up to a 40% increase in the optical power by employing a nano-lithography technique MOCVD [TOH, CHE 08]. The main reason for this improvement is that such patterns help reduce the stresses on the GaN/sapphire interface leading to a reduction in the defect density (dislocations) of the sapphire substrate to less than 10^8 cm^{-2}. The result of improving the quality of the substrate is a better thermal dissipation and the increase in optical power for an equal given supply current (20 mA).

The hemispherical pattern (Figure 1.12(c)) is more effective, since by spacing the microspheres 5 µm we obtain an increase in the optical power of 155% compared with conventional LEDs (Figure 1.12(a)) [OH 08]. It is also shown that the reduction in the dislocation density (up to 10^8 cm^{-2}) is a function of the space between the various patterns obtained by ICP dry engraving and photolithography. This also leads to better heat dissipation. Finally, the technique using a pyramidal pattern (Figure 1.12(d)) proves to be very effective and improves the optical power by 25–70% [WUU 06, HUA 09, LEE 07].

The large difference between the refractive indices of air and GaN (n = 2.85) has led to numerous studies on the indium oxide contact layer doped with tin (ITO – Indium Tin Oxide) to obtain transparent contact and avoid the phenomenon of total internal reflection in the GaN chip. The ITO was chosen for its high conductivity (105 Ω^{-1}.cm^{-1}), its high optical transmittance (90%) and absorbance at 420 nm (α = 664 cm^{-1}) generally lower than most of the thin metal films ($\alpha = 3.10^5$ cm^{-1}) [RAY 83, MAR 99]. An ITO layer is obtained by cathode radio-frequency pulverization. Annealing in situ improves its transmittance by up to 97%, thereby reducing the operating voltage from 5.74 to 4.28 V [CHA 03]. Another study showed that the ITO could also be used on power LEDs, thus leading to an increase in external efficiency of 46% and an improvement of the optical power of 36% [CHA 05]. The most innovative way to improve light extraction through the ITO material is to produce a GRIN (graded refractiveindex-ITO – GRIN-ITO) and anti-reflection (AR) ITO layer. The luminescence efficiency improves by 24.3% compared with an LED with a solid ITO layer (n_{ito} = 2.19). This is due to the sharp reduction in the Fresnel reflection in the ITO/air interface (1.17 $<n_{GRIN\ ITO}$ $<$2.19) [KIM 08]. The phenomenon of total internal reflection in the GaN chip represents a fourth obstacle. Other proposals have contributed to the improvement of the optical power. Structuring of photonic crystals, for example, has been applied to the GaN contact P-layer.

Figure 1.12(e) shows a diagram of a GaN LED with photonic crystals (PC LEDL) [NG 08]. Several studies have proven the effectiveness of inserting photonic crystals, demonstrating that the optical power can increase by up to 40% [LIU 09, DAV 06] or almost triple compared with conventional structures (Figure 1.12(a)) [BYE 07]. Photonic crystals are usually made by nano-lithography or holographic laser [NG 08, LEE 09].

An active region based on quantum dots 3 nm high and 10 nm wide can be developed. By injecting currents of 3–50 mA, the gap of the LED's active region drifts by 68.4 meV. This drift is very low due to the insensitivity of quantum dots to current and temperature. This device is therefore stable in wavelength despite the auto-heating phenomenon [SU 04]. Indeed, the energy levels in a quantum dot are very weakly dependent on temperature. This result is important for applications where the environment's temperature is changing significantly: automotive, space, aeronautics, etc.

The basic problem of the configuration presented so far is the compromise between the extraction of light and the uniformity of the injection of current over the entire surface of the chip. In consideration of the increase in the injection current, increasing the thickness of the semi-transparent contact, necessary to avoid excessive heating, reduces the amount of light extracted from the chip by light absorption. In the flip chip configuration (Figure 1.13), a reflective metal layer ensures, on the one hand, the P contact of a large part of the diode and, on the other hand, increases the extraction efficiency by the reflective character of this metal layer [CHE 08, CHO 06]. In this architecture, the light is emitted through the sapphire substrate.

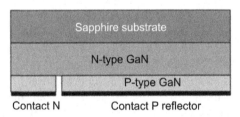

Contact N   Contact P reflector

Figure 1.13. *FC-type flip chip structure diagram*

Moreover, the structure is transferred to a material that has a low thermal resistance, such as silicon or certain ceramics such as Al_2O_3 or AlN. The reconciliation of the junction and the thermal dissipater provides better heat management. This configuration is often called P- down type, indicating that the P layer is attached to the substrates. We thus significantly reduce the thermal resistance of the device.

Finally, besides the advantage from the thermal point of view, this architecture eliminates the presence of superior bonding wires (N contact) and potentially facilitates conformal deposition of the phosphor light conversion [CHE 08, ZHE 14, DAV 14].

In the case of a vertical thin film configuration (Figure 1.14), the P contact is deposited on the entire surface of the LED chip after the GaN epitaxial growth [HAN 15, CHU 13]. The structure is then transferred onto a host substrate. The next step is to remove the substrate to open the N contact. This can be done by a method consisting of removing the substrate by laser (Laser Lift Off, LLO).

In the VTF configuration, the N contact is deposited in the center of the chip and the low electrical resistivity of the N-type material is low enough to ensure a good flow of current across the surface of the LED chip.

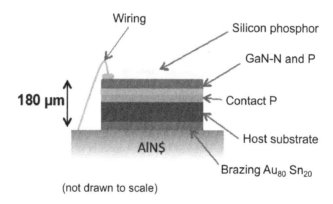

Figure 1.14. *VTF-type vertical thin film structure diagram*

This configuration allows for good light extraction through the GaN-N but makes it difficult to deposit phosphors through the presence of superior bonding wires. Companies such as CREE, OSRAM or Semileds offer this chip configuration in their products.

The thin film flip chip solution (TFCC) shown in Figure 1.15 corresponds to a combination of FC and VTF configurations combining the benefits in terms of heat dissipation, the lack of wired cabling and eased deposition of phosphors [DIN 15]. The solution consists of removing the substrate according to the process once the plate is realized according to the FC configuration. It is in fact the natural evolution of the FC solution. The light is thus extracted through the GaN-N.

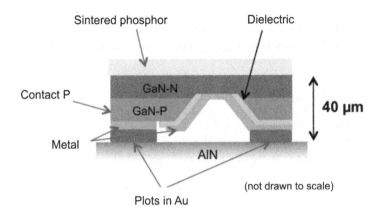

Figure 1.15. *TFCC-type thin-film flip chip structure diagram*

1.2.2.2. *State-of-the-art GaN LED assemblies*

The main role of the assembly is to make a component usable in its final environment. An assembled optoelectronic component has two parts allowing an external exchange: an electrical and an optical part.

The electrical part is relatively conventional and is similar to what can be seen in electronic devices. We thus find DIL (Dual In Line) or CMS (Surface Mounted Components)-type connections as appropriate. The different connectors carry either the power or the signal to be processed. In the case of a GaN LED for public lighting, there are mainly two power connectors.

The optical part is used to format the optical emission; it is a photonic device. Depending on the application, there is focalization, polarization or change by nonlinear optical frequency functions. Once again, in public lighting, we find the frequency change by nonlinear optical function (photoinduced emission) in the assembly materials.

The use of the component in its final environment requires specific protection against different physical and chemical stresses. The casing will be different according to US standards (MIL-STD) or Japanese ones (JIS and JEITA), intended to isolate the bare chip from different environmental stresses. Table 1.7 summarizes the various American and Japanese standards depending on the environmental stresses for low- and high-power GaN LEDs (up to 3 W).

Type of test	Associated standards		
	United States	Japan	
	MIL-STD	JEITA	JIS
Thermal cycles	202 : 107D 750 : 1051 883 : 1010	ED-4701 (100, 105)	C 7021 : A – 4
Thermal choc	202 : 107D 750 : 1051 883 : 1011	–	–
High-temperature cycles	–	ED-4701 (200, 203)	–
High-temperature storage	883 : 1008	ED-4701 (200, 201)	C 7021 : B – 10
Low-temperature storage	–	ED-4701 (200, 202)	C 7021 : B – 12
Temperature/humidity storage	202 : 103B	ED-4701 (100, 103)	C 7021 : B – 11
Resistance to the brazing temperature	202 : 201A 750 : 2031	ED-4701 (300, 302)	C 7021 : A – 1
Brazing thermal resistance	202 : 208D 750 : 2026 883 : 2003	ED-4701 (300, 303)	C 7021 : A – 2
Electrostatic discharge (ESD)	–	ED-4701 (300, 304)	–
Vibrations	–	ED-4701 (400, 403)	–

Table 1.7. *American and Japanese high- and low-power GaN LED qualification test standards*

Most qualification standards in the field of GaN LEDs take temperature into account. This is explained by the auto-heating phenomenon involving a change in the junction temperature inducing a drift in the central wavelength of the LED and therefore a color change. The two major issues identified thus far are temperature and light extraction. This has created great interest in two major parts of encapsulation: the encapsulating gel and the base serving as a thermal dissipater.

Thermal management is one of the most important factors in the encapsulation of an LED. This is all the more important now that the power density of LEDs is becoming more consistent (P_{opt} > 100 mW) as well as the requirements in terms of reliability (λ > 50,000 h). Figure 1.16 shows the encapsulation of low-power "5 mm"-type LEDs (<30 mW).

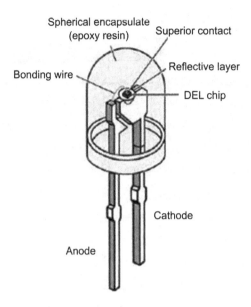

Figure 1.16. *"5 mm"-type classic encapsulation*

This classic encapsulation is that used in the 1970s. The chip, reflective layer and the electrical connections are embedded in an epoxy resin serving as a collimating lens and material promoting light extraction. Casings of diameter 5 mm of this period were mainly used for low-power LEDs and had a thermal resistance of 250 K.W^{-1}. Figure 1.17 presents the evolution of the thermal resistance for different types of casings from 1970 untiltoday [SCH 06].

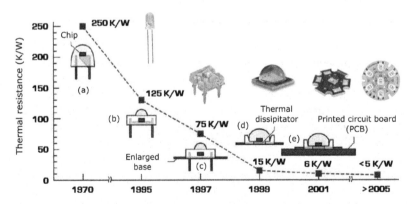

Figure 1.17. *Thermal resistance from 1970 to 2005: a) 5 mm casing, b) lowered casing, c) lowered casing with enlarged base, d) power casing with thermal dissipaterand e) power casing on printed circuit board (PCB)*

Encapsulation casings of the 1990s separated the base from the collimating lens to reduce the thermal resistance of up to 75 K.W^{-1}. In the early 2000s, many power LEDs were equipped with "Barracuda"-type casings (Philips Lumileds) using an aluminum or copper thermal dissipater as the base of the chip, thereby reducing the thermal resistance to 15 K.W^{-1}. By mounting the latter on a PCB base, the thermal resistance was reduced to 6 K.W^{-1}. Today, the bases are PCBs in the shape of stars (REBEL STAR) of ceramic, copper or aluminum. This enables the technological advantage of placing up to seven chips on the same base. These technologies are mainly used for white LEDs power (> 100 mW).

To exceed optical powers of 1 W or efficiencies of 180 lm.W^{-1}, the method used is Chip On Board (COB), widely used in hybrid thick layers (1980) in microelectronics. This allows manufacturers to offer LED components containing at least four LED chips and up to 128 [WU 12]. Figure 3.3 provides a COM-type or MCM (multi-chip module) component.

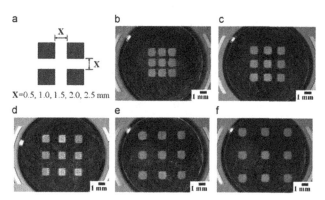

Figure 1.18. *Positioning of GaN chips on the board: a) theoretical positioning, b) type A (space of 0.5 mm), c) type B (space of 1.0 mm), d) type C (space of 1.5 mm), e) type D (space of 2.0 mm) and (f) type E (space of 2.5 mm)*

One of the most effective ways to obtain white light is to use a layer of yttrium aluminum garnets doped with cerium (YAG: Ce)-based phosphor that converts blue light of a GaN LED into white light. Thus, many studies have been conducted on the encapsulation of these LEDs for the management of heat dissipation. Shin *et al.* compared three silver thermal pinholesdesigns for power GaN LEDs. Figure 1.19 shows these thermal pinholes [SHI 06].

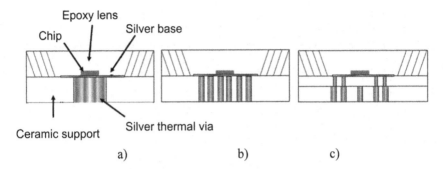

Figure 1.19. *Design of thermal pinholes for power GaN LEDs: a) Unique via with a diameter of 1.47 mm, b) 16 vias with diameters of 0.43 mm and c) two layers of pinholes: superior, 16 pinholes with diameters of 0.26 mm; inferior, ninepinholes with diameters of 0.5 mm*

This study showed that the most effective model to dissipate heat is through the unique via as shown in Figure 1.19(a) with an optical power of 1.139 W (2% higher than the other two) and a thermal resistance of 48.9 K.W^{-1} (15–20% lower than the other two). Another design is possible by creating copper thermal pinholes of 0.8 mm diameter through a dielectric insulating layer FR4 (FR-4: Type4 Retardant Flame) to reduce the thermal resistance of the assembly to 4 K.W^{-1}[CRE 10].

A second study conducted by Fan *et al.* has demonstrated that the insertion of a heat insulating layer, placed between the chip and the phosphor layer reduces the diffusion of the junction temperature of the chip in the layer of phosphor. Figure 1.20 shows a diagram of the technology put in place within the framework of this study [FAN 07].

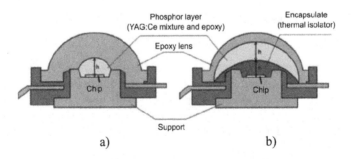

Figure 1.20. *Cross-section of white power LEDs: a) conventional white LED and b) thermally isolated white phosphor LED*

At 500 mA, the junction temperature of the chip is 125°C. The temperature of the phosphor layer was reduced by 16.8°C compared with that of the conventional LED. The need to reduce the temperature of the epoxy/phosphor mixture is directly related to the fact that many coatings are sensitive to thermal stress.

The heat dissipation of LEDs or COB power LEDs (> 1W) requires a heat dissipation study that is correlated with infrared-type measures, T3STER or I (V) pulse-type measures, etc.

Figure 1.21 shows an example of results obtained by finite element method (FEM) on a COB structure containing 25 chips [SHI 15].

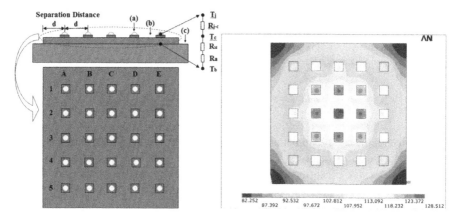

Figure 1.21. *Isotherms obtained by FEM simulation on a COB system for a consumption of 1.5 A. For a color version of the figure, see www.iste.co.uk/deshayes/reliability2.zip*

The distance between each LED is 10 mm with a set transferred onto a PCB of 6 × 6 cm. The simulation indicates that the lateral heat flow is significant and has an impact on the maximum temperature of the device mainly in its center. We thus find results that are well-known in microelectronics. One of the methods, well known for determining precisely the heat flow of a device, is the use of a pulse measuring system (example shown in Figure T3STER 1.22). The parameter to follow is the voltage of the diode, which may be compared with the temperature by a k linear relationship factor. The system can therefore determine the junction temperature for a given polarization. We can deduce, subsequently, the equivalent thermal resistance of the device by a simple relationship between the temperature, power dissipation and thermal resistance [YAN 12].

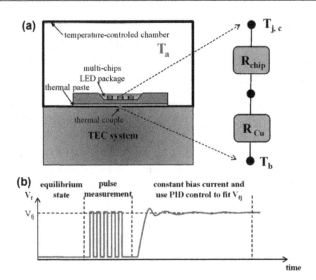

Figure 1.22. *Study of a COB pulse measuring device*

This technique can be implemented with a pulse current generator (modular laser power + a pulse generator) associated with an oscilloscope. The frequency performance is not exceptional since the LED's cavity (active area) is relatively large and therefore requires pulses in the order of a few tens of microseconds to a few hundred microseconds. A compromise on the duration of the pulse (negative) involves reverse recovery time (extinction time at the source) coupled with a pulse duration suited for not changing the device's temperature during the voltage measurement across the LED terminals.

A second device can perform the same measurement but with more comprehensive information on the thermal flow structure. The T3STER determines the thermal capacity depending on the thermal resistance. We can, according to a precise protocol, determine all the layers composing the assembly.

We use coating gels to reduce reflection losses and absorption, and total internal reflection at the chip/air interface. The quality and sustainability of this gel's performance must be guaranteed in order to optimize device performance during its lifetime. The coating gels have seen many improvements over the last 20 years. The coatings most frequently used until the early 2000s were epoxy resins. For many years, they were chosen for their high refractive index (1.4–1.6) [LIN 10], good thermal stability up to 120°C [SCH 06, HEL 08] and high transmittance (85–90%)

in the range of 500–800 nm [SIL 07]. However, many problems were highlighted with the growth of GaN and power LEDs. Yellowing of the resins was observed when subjected to radiation in the blue and UV range. This phenomenon contributed to the development of new chemical formulations of the epoxy resins that were polymerized by UV radiation so as to optimize their resistance to high temperatures (>120°C). Kumar *et al.* have reported glass transition temperatures between 140 and 180°C, thus enhancing the light extraction of the LED [KUM 06]. Another study showed that, by controlling the current (1–30 mA) and the time (1–4 s) of polymerization of photosensitive epoxy resins, the profile of the output beam, the luminous intensity and the angle of emission of the LED could be improved [WAN 08]. Indeed, the emission angle was reduced to 40° and the light intensity was increased by a factor of 3. Furthermore, with the increase in the optical power, and therefore, the increase in the temperature junction, the same phenomenon of yellowing is seen in temperatures above 120°C [SCH 06].

To address these problems, epoxy resins have been replaced by silicon coatings. These silicon oils have been chosen for their high transmittance (> 95%) in the blue range [SIL 07], their excellent thermal stability (>180°C) [SCH 06, SIL 07], their high refractive index (1.4 – 1.6) [RIE 04], their low optical losses of 400 – 1,310 nm (<0.04 dB.cm^{-1}) [JAE 07] and homogeneity, avoiding optical losses due to Rayleigh scattering [DEG 04]. These are usually methyl- and phenyl-based polymers such as PDMS (poly dimethyl siloxane) or PMPS (poly methyl phenyl siloxane) or even PMPS/PDMS copolymers. Table 1.8 summarizes the optical losses of these three types of polymers at different wavelengths [JAE 07].

Type of polymer	Optical losses (dB.cm^{-1})					
	1,550 nm	1,310 nm	850 nm	633 nm	400 nm	300 nm
PDMS	0.67	0.14	< 0.01	< 0.01	0.03	0.09
PMPS	0.62	0.35	< 0.01	< 0.01	< 0.01	0.55
PMPS/PDMS	0.66	0.28	0.03	0.03	0.04	0.24

Table 1.8. *Optical losses of the three main silicon oils used in LEDs at different wavelengths*

Their importance in the UV and the IR ranges is also seen by their transmittance. Figure 1.23 compares epoxy resin and silicon oil in an aging test in transmittance temperature in terms of the wavelength [SIL 07].

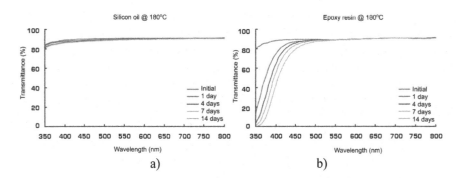

Figure 1.23. *Aging test at 180°C of the transmittance as a function of the wavelength: a) silicon oil and b) epoxy resin. For a color version of the figure, see www.iste.co.uk/deshayes/reliability2.zip*

Silicon oils are also very elastic (<0.5 GPa), which causes less mechanical stress on the chip and the bonding wires [WON 89].

Although silicon oils have better thermal properties and better resistance to blue radiation than epoxy resins, they are susceptible to photonic energy radiation higher than 3 eV. Indeed, Koizumit *et al.* have demonstrated that UV radiation causes electronic transitions in the σ methyl bonds in the PDMS. Similarly, in the PMPS, electron transitions, occurring at the π bonds of the phenyl and, at higher energies, σ methyl bonds are responsible for the fluorescence emission in UV [KOI 92]. This has led to the recent development of silicon oil consisting of a mixture of vinyl PDMS and a hydromethyl siloxane/PDMS copolymer. Lin *et al.* have therefore compared a high-transmittance silicon oil, optical gradient epoxy resin and new improved silicon oil. After 460 h of aging at 120°C under UV radiation, the improved silicon oil lost only 1% of its transmittance unlike the normal oil (20%) and the optical gradient epoxy resin (> 40%) [LIN 10].

Other coatings have been proposed to improve extraction of light and reduce the Fresnel reflection at the chip/air interface. The addition of nanoparticles such as yttrium, zirconium, GaN, SiN, AlN, ZnSe or ZnS obtains a high refractive index (2.3 – 2.9) and contributes to the improvement of light extraction at the LED's output [LES 98]. The integration of mineral diffusers such as TiO_2, CaF_2, SiO_2, $CaCO_3$ or $BaSO_4$ is a technique used to standardize the output beam of multichip LEDs [REE 06]. The superposition of several layers of different refractive indices improves the external luminescence efficiency by up to 40%, the outer layer having the lowest optical index [LEE 04].

Finally, regarding the technology of white power LEDs, the addition of phosphor for converting blue light emitted by chip into white light has been studied extensively. Different chemical compositions of the phosphor and the silicon oil, as well as assembly geometries, have been tested over the last decade. The most common chemical composition is that of the YAG:Ce phosphor based on the yttrium aluminum garnet ($Y_3Al_5O_{12}$) doped with cerium (Ce3 +). With an excitation wavelength at 460 nm from the blue LED, the $5d^1 \rightarrow 4f^7$ electronic transitions of the phosphor Ce^{3+} in the sites of $Y_3Al_5O_{12}$ allow us to obtain an emission band in green/red (540 nm) [ZHA 08]. When this phosphor is in direct contact (pelleted) with the chip, an optical power loss by absorption of 60% is observed [NAR 05]. When the latter is separated and positioned above using smooth (Ag) reflective layers, the luminous efficiency is improved by 36%. Using diffuse reflective layers, efficiency is improved by 75% [KIM 05b]. Luo *et al.* demonstrated that the addition of a spherical lens above the phosphor layer increases the light output by 20% [LUO 05].

A technological barrier was lifted by working on the color yield index. Indeed, the IRC of a white LED using the phosphor YAG: Ce, is usually around 75. Won *et al.* [WON 89] therefore proposed the use, always from a blue LED, of two phosphors separated by a layer of silicon oil: the phosphor (Ba, Sr) $2SiO_4$: Eu^{2+} green and $CaAlSiN_3$: Eu^{2+} red. With an excitation wavelength at 458 nm from the blue LED, the $5d^1 \rightarrow 4f^7$ electronic transitions of the phosphors Eu^{2+} in the sites of $(Ba, Sr)_2SiO_4$ and Ca of $CaAlSiN_3$ allow one to obtain emissions in green (525 nm) and red (637 nm), respectively. This has led to a CRI of 95 and a luminous efficacy of 51 $lm.W^{-1}$ at 350 mA [CRE]. The same study showed the emission of white light with the following two phosphors (Ba, $Sr)_2SiO_4$: Eu^{2+} green (525 nm emission band) and Sr_3SiO_5: Eu^{2+} orange (emission band at 610 nm). This structure has achieved a CRI of 83 with a light output of 55 $lm.W^{-1}$ at 350 mA.

1.2.2.3. *Structures of studied GaN LEDs*

In this book, two types of components will be studied:

– A component using a conventional TO47 casing with lens mounting and a diameter of "5 mm" with a low-power light emission (<30 mW) on GaN. We will show the impact of the central wavelength of emission on the drift of the elements of the encapsulation casing;

– A star MCPCB high-power-type structure (> 100 mW) GaN technology with YAG: Ce phosphor.

Figure 1.24 shows a simplified diagram of the studied GaN LEDs.

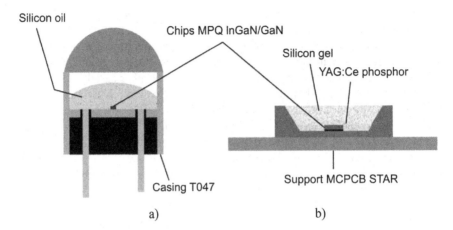

Figure 1.24. *Simplified diagram of the studied structures: a) GaAs and GaN LEDs on T047 casings and b) white GaN LEDs on an MCPCB star base*

Low-power LEDs are components sold by the Japanese company OPTRANS. The GaN technology is a multi-quantum well InGaN/GaN structure with two upper contacts on a sapphire substrate. The encapsulation of these two technologies is the same and is specially designed for the field of space. The casing is T047 Kovar and is resistant to solar radiation. The chip and encapsulation are described in Chapter 3.

White GaNpower LEDs are multi-quantum well InGaN/GaN-based vertical structures on a Si substrate, where the conversion of light is performed by a silicon oil mixture and sediment YaG phosphor. The chip is soldered onto a MCPCB STAR star-shaped base. The chip and encapsulation are both described in Chapter 4 of this book.

1.3. Positioning, justification and objectives of the study

State-of-the-art of GaN technologies endeavoring to describe both the physical properties of nitride materials, structures, components and assemblies associated with low- and high-power LEDs has been achieved. This has enabled the description of the technologies discussed in this book. At this stage, we should describe the positioning of our study within a national and international perspective.

1.3.1. *Positioning and justification of the study*

Internationally, the great challenge of research and development of GaN technologies is how to obtain higher reliability at a lower cost. However, today, many qualification standard tests require a high number of components. Table 1.9 groups the qualifying standard tests used by CREE (USA) and Ledman (China) on low-power (<0.1 W) and high-power LEDs (0.1–3 W) [CRE, LED].

Test	Standard	Test conditions	Duration	Number of LEDs
Thermal cycles	JEITA ED-4701 100 105	–40°C/25°C /100°C/25°C 30 min/5 min/30 min/5 min	100 cycles	f: 100 F: 30
Choc cycles	MIL-STD-202G	–40°C/100°C 15 min/15 min (Ledman) 30 min/30 min (CREE)	f: 100 cycles F: 300 cycles	f: 100 F: 30
High-temperature cycles	JEITA ED-4701 200 203	25°C/65°C RH: 90%, 24 h for 1 cycle	f: 10 cycles F: 50 cycles	f: 100 F: 30
High-temperature storage	JEITA ED-4701 200 201	$T_{ambient} = 100°C$	1,000 h	f: 100 F: 30
Low-temperature storage	JEITA ED-4701 200 202	$T_{ambient} = -40°C$	1,000 h	f: 100 F: 30
Temperature/hum idity storage	JEITA ED-4701 100 103	$T_{ambient} = 60°C$, RH: 90%	1,000 h	f : 100 F : 30

Table 1.9. *Environmental qualification standards for low- (f) and high-power (F) GaN LEDs*

RH: relative humidity

OSRAM enterprises (Germany), Nichia and Toyoda Gosei (Japan), Philips Lighting (USA), Seoul Semiconductor (Korea) or Everlight (Taiwan) use these qualification standards to assess their LED lamp technology. The duration of these aging cycles can also be very costly. Some research laboratories (Nanyang Technological University, Singapore [MIN], Pacific Northwest National Laboratories, USA [RIC 10] and industrial GE Lighting, USA [SMI 10]) recommend testing up to 6,000 h. This makes if possible to predict the lifetime of the components thanks to mathematical laws outlined in the first part of this chapter.

The use of accelerated tests can limit the number of hours of aging to 1,000 h. Table 1.10 shows the accelerated testing standards used for GaN LEDs [CRE, LED].

Test	Test conditions	Duration	Number of LEDs
Standard active storage	$T_{ambient} = 25°C$ f: I = 30 mA (blue and green), 50 mA (red) F: I = 700 mA	1,000 h	f: 100 F: 30
Active storage + humidity	$T_{ambient} = 60°C$, RH: 90% f: I = 20 mA F: I = 500 mA	f: 500 h F: 1,000 h	f: 100 F: 30
Low-temperature active storage	$T_{ambient} = -30°C$ f: I = 20 mA F: I = 600 mA	1,000 h	f : 100 F : 30

Table 1.10. *Standard accelerated tests*
for low- (f) and high-power (F) GaN LEDs

In the laboratory, other methods based on failure analysis using electrical and optical non-destructive analyses explain the mechanism of degradation induced by accelerated tests. Meneghini *et al.* (University of Padova, Italy) dissociates, for example, the effect of temperature and current from the electrical characteristics (current–voltage I (V), capacitance–voltage C (V)) and optical (optical spectrum L(E) and optical power P(I)) of GaN LEDs [MEN 07, MEN 06b]. This method, being non-destructive, makes it possible to dissociate the component failure from that of the assembly [TRE 07].

In France, if the lighting sector is broken down into its various components, material, substrate, epitaxy and process manufacturing, chip packaging, lighting and finally uses (building, public lighting, etc.), the French industrial panorama covers mainly the sectors of lighting fixtures and accessories (Philips, Lenoir, EcceLectro, Petzl, Flux lighting, Gaggione). There are also a number of laboratories dedicated to the study and development of substrate materials and heterostructures used for light emission (CEA-LETI, CHREA). There exists therefore in France, in the value chain of this area, a significant gap between the material's industrials and R&D on the shaping of the materials and the lighting application area. However, this is not the case at European level where OSRAM as PHILIPS are, for this sector, vertically integrated companies from the chip to the lighting device [DUB 09]. This panorama lets us say that few French actors are involved in the reliability and failure analysis of power LEDs.

In the IMS laboratory (Bordeaux), the EDMINA team has implemented several methodologies since the early 2000s for the evaluation of reliability and failure analysis of optoelectronic components. The design of reliability from the point of view of the EDMINA team is represented by Figure 1.25.

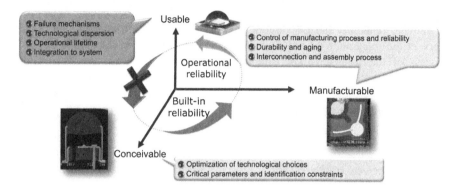

Figure 1.25. *Conception of reliability by the EDMINA team*

This design revolves around the three steps leading to the manufacture of a component: the conceivable, manufacturable and the usable. Generally, the construction of reliability is established from the area of the manufacturable up to the usable. The manufacturable is an area in which reliability can be constructed to help control manufacturing processes, evaluation of technological sustainability and the assembly part. The usable is the final step of reliability because the component is assembled and operational. The construction of reliability here lies in the assessment of failure mechanisms under environmental stresses and/or accelerated aging tests, the evaluation of technological dispersion, the determination of the operational life of the component by methodologies based on physics and, finally, the integration into the complete system. Several theories have therefore been carried out at this stage of reliability construction, called "operational", with different approaches: failure analysis and signatures [DES 02], failure analysis and physical simulations [HUY 05] or statistics and component/system interaction [MEN 06a].

Since 2008, the concept of the EDMINA team is to bring the concept of reliability to the earliest design phase of the components (conceivable), as well as the development phases. This will include helping to establish adapted and optimized technological choices (materials, interfaces, assembly), and to identify the critical parameters and environmental stresses linked to the application. This framework is called reliability of the "constructed".

Our study is a methodological approach of failure analysis using the two types of reliability construction: constructed reliability (public lighting, power LEDs) and operational reliability (spatial, low-power LEDs). The second added value of this work is the integration of physical and chemical analyses in the failure analysis process of the studied LEDs. The latter will help to confirm the degradation mechanisms previously localized by electrical and optical default signatures extracted from electro-optical non-destructive measurements. Another advantage of this methodology, and the contribution of physicochemical analyses, is that it does not require a high number of components of which the maximum is 4 per technology.

1.3.2. Objectives of the study

In a favorable international environment and a growing market, extending the lifespan of white GaN-based power LEDs is a key factor. Today, many manufacturers offer products whose lifespan is over 10,000 hours. The failure criterion, defined relative to the flow (lumen), is generally specified to limits (L→ Lumen maintenance) corresponding to a loss of 50% (L50), 30% (L70), 20% (L80) and 10% (L90) of flux according to the public lighting standards.

Table 1.11 shows the results of aging tests at the temperature of LED lamps manufactured by GE Lighting[SMI 10].

Test conditions	Flux at 6,000 h	Exponential projection[a]
$T_{ambient} = 25°C$	96.1%	L85 > 50,000 h
$T_{ambient} = 47°C$	96.0%	L70 > 50,000 h L85 > 25,000 h
$T_{ambient} = 60°C$	93.3%	L70 ≈ 49,000 h

Table 1.11. *Real and exponential projection of the flux in operational conditions at three ambient temperatures*

[a]L85 and L70 represent the limits at 85 and 70% of flux respectively.

Figure 1.26 summarizes the aging results with the exponential projection [SMI 10].

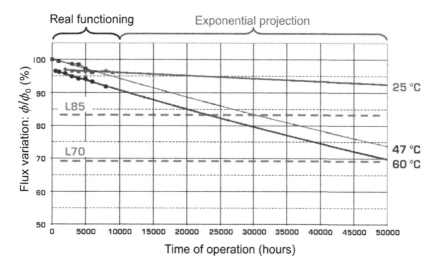

Figure 1.26. *Flux variation as a function of the LED lamp's operating time*

The law of the decrease in flux, φ, in time is also exponential and can be expressed by equation [1.1].

$$\frac{\varphi}{\varphi_0} \; (\%) = \exp\left(-\frac{t}{\tau}\right) \qquad\qquad [1.1]$$

τ is the degradation factor (homogeneous at one time).

At 25°C for 6,000 hours, LED lamps have lost nearly 4% of their original flow. If it is based on a mathematical projection using an exponential distribution, the life displayed by the manufacturer is greater than 50,000 h (Table 1.11) for 15% flux loss (L85). Equation [1.2] shows the empirical Arrhenius law frequently used to determine a temperature acceleration law.

$$A_T = \exp\left[\frac{E_a}{k}\left(\frac{1}{T_{J\,@\,25\,°C}} - \frac{1}{T_{acc}}\right)\right] \qquad\qquad [1.2]$$

A_T, acceleration factor; E_a, activation energy; k, Boltzmann's constant; $T_{J\,@25°C}$, junction temperature at $T_p = 300$ K and T_{acc}, temperature in accelerated conditions.

– The acceleration factor for L85 and $T_J = 25°C$ is about 1.5 for $T_{acc} = 47°C$ and 1.9 for $T_{acc} = 60°C$. The activation energy is estimated to be 158 meV, which makes if possible to calculate a lifespan of about 10^5h at L85 and T_J = 25°C. The acceleration factors are generally weak in the case of assembled components. An activation energy constant, whatever the temperature acceleration, means that it is assumed that a single mechanism of degradation occurs. However, the interaction of the chip with its assembly can generate several failure mechanisms simultaneously. The Arrhenius law is therefore better adapted to "bare chip" components aging. Its application on assembled devices requires knowledge of all failure mechanisms involved in the degradation of the component.

The problem is that these laws work only under two conditions: the environmental constraint is the temperature and the chip is "bare". Otherwise, the law of acceleration is much more complex and it is very difficult to extract the lifetime in operational conditions [FUK 91, UED 96].

Mathematics is often used to estimate the degradation law and extrapolate lifetime. When these laws are set by at least two coefficients, we can build a correlation law to simulate the corresponding points in the first moments of failure. For that, empirical methods are used (e.g. AT^M law) and/or statistics, the Monte Carlo method, Bayesian probabilities or Weibull distribution [FUK 91]. The fundamental problem in this methodology comes from the mathematical breakdown of the law. Indeed, it does not follow a physical law and therefore cannot describe the internal component failure phenomena. Extrapolations may even become false in the case of degradation from several physical phenomena. In these cases, we can give an optimistic or pessimistic result on the distribution of lifespan of the studied technology.

Extrapolation of the lifespan, by mathematical laws, remains a method where the confidence index is often less than 70%. To improve this critical parameter in the estimation of reliability, we develop failure analysis by physical methods. The main objective of our work is to build a methodology that is based on physics to model failure analysis of GaN LEDs. The aim is to explain, from the electrical and optical failure signatures, the main physical mechanisms involved in the degradation of the component. When the device is still young, failure analysis allows us to establish the robustness and quality of the technology. In this case, the physical analysis provides design assistance and thus builds upon the reliability design. For a mature and marketed device failure, analysis allows building a law of degradation and identifying failure mechanisms. On a more mathematical note, we can construct an algorithm for estimating the distribution of lifespan. In this case, the physical analysis provides support for the technological choice of a component based on the mission and environmental stresses. As part of this book, these two aspects have been developed [DES 02]. This study particularly focuses on physical failure analysis through two projects:

– application of the methodology to a commercial device (low-power InGaN/GaN LED). It corresponds to the concept of operational reliability, classic reliability construction pattern (COTS component →Commercial off-the-shelf). This part is the subject of Chapter 3;

– application of the methodology for industrial applications of white power GaN-based LED of the manufacturer CREE. This demonstrates that the methodology can meet industrial challenges by finalizing the study with proposed technological solutions (Chapter 4). LEDs are used here for public lighting. The collaboration between the outlook of the EDMINA team: the integration of reliability from the component design stage (built reliability).

From electrical and optical models adapted to the studied components, we can extract the material parameters in order to pre-locate/the failure(s) involved in the degradation of the device. The specificity of this methodology lies in the integration of physical and chemical analyses to confirm the degradation mechanisms.

In both types of studies, several important issues caught our attention:

– what are the types of critical aging for these components? Are they the same for low- and high-power applications? Different types of aging having temperature, current and radiation as constraints were performed. Active storage ($1,500$ h/T_{MAX}/I_{rated}) proved to be the major contributing factor for both studies. The results presented in this research will therefore focus mainly on the deterioration in active storage;

– what part of the degradation should be allocated to the bare chip component and to the assembly? Are they distributed differently according to the power? It appears that the assembly is the critical part of the current technologies of GaN LEDs, especially for low-power ones. For example, coating the chip with silicone gel in each technology proves to be a critical point of the assembly when subjected to blue/UV radiation. This phenomenon increases with temperature and with power (white LEDs). We may consider a photo-thermal reaction of the polymer and therefore a degradation of its optical properties;

– what are the means put in place to effectively identify critical points and degraded areas? The faulty zones appear to be localized in the component part assembly, and more specifically, in the polymer coating. To highlight, nondestructively, the degradation phenomena, electro-optical characterizations are used. These make it possible to pre-locate the fault, and thus, establish the physicochemical analyses necessary and appropriate to confirm the degraded area.

1.4. Conclusion

The aim of this chapter is to present state-of-the-art GaN technologies by focusing on three main points: the market for Leds and GaN technologies, nitrides materials and low- and high-power GaN components with associated assemblies. This chapter synthetically presented the evolution of these technologies leading to developments that are increasingly complex and precise for uses that are increasingly important. Today, there are LED lamps in local stores at prices that are still a little high but with guaranteed lifetimes of more than 10,000 h.

The synthesis of all GaN technologies has allowed us to position our study nationally and internationally, and to compare with the work and concepts developed by the EDMINA research team:

– the qualification standards are frequently used by international manufacturers (OSRAM, Nichia, Cree, Philips Lumileds, Ledman, etc.);

– statistical and mathematical methods make it possible to predict the operational lifespan from the integrated component to the system;

– the failure signatures approaches, physical simulations and statistics were developed by the EDMINA team;

– integration of reliability at the component design stage helps answer fundamental challenges facing industry.

The chosen approach is that of a methodology based on failure analysis and extraction of electrical and optical signatures to locate the fault. Adding physicochemical analysis reduces the number of components to study and offers the opportunity to confirm the degradation mechanisms induced by aging in active storage. Our study aims to build reliability across multiple criteria such as pre-assessment and durability of materials, and capitalization of reliability data.

The objectives of the next chapter are to develop the foundations for this study by describing the electrical and optical mechanisms involved in the GaN-based LED technology, describe electro-optical characterization means developed at the IMS Laboratory, and to present the objectives of physicochemical analysis used in this research.

2

Tools and Analysis
Methods of Encapsulated Leds

Optoelectronic assemblies have greatly evolved over the past 15 years. The factors responsible for this evolution are mainly emerging applications such as public lighting, mastering new technologies such as GaN and the cost that is reasonable for general public use. Cost reduction involves, in general, the increase in the amount of plastics in the assembly process, replacing glass, ceramics and solders. Indeed, synthesis, manufacturing and shaping are less expensive for the plastic materials than for materials such as glass or ceramics. This technological breakthrough took off at the beginning of this millennium and many plastic materials were introduced in the optoelectronics industry in many forms:

– adhesives for maintaining the chip on its support;

– embedding of the chip to improve the optical efficiency;

– plastic lens.

This list, being non-exhaustive, makes the analysis of an assembled component structure much more difficult. Indeed, the analysis of plastic materials, consisting generally of relatively long and fragile molecular chains, can hardly be recognized as complete with conventional physicochemical analyses used in optoelectronics (SEM, EDX, SIMS, etc.). During this research, a number of additional analyses were therefore carried out in collaboration with chemistry laboratories of the University of Bordeaux. This activity is a major focus of this research.

Chapter 1 highlighted the fact that the technology used mainly for public lighting is of GaN type. This technology, still in development, required physicochemical analyses compatible with the semiconductor in order to determine the chip's structure. These analyses (SEM (Scanning Electron Microscopy), EDX, SIMS, RBS and PIXE) have helped in constructing a physical model justified by measurements on the structure and materials. The electrical and optical measurements carried out on the studied LEDs yielded quite different results from the analyses already made within the EDMINA team on GaAs technologies. Chapter 3 will show that the contacts are responsible, to a great extent, of the type of characteristics observed by electrical analyses. For optical analyses, more particularly optical spectra, additional emission patterns of electroluminescence have been demonstrated. It will be demonstrated in Chapter 3 that the polymer coating the chip interacts with the light emitted by the LED (464 nm) to give a fluorescence-type emission.

Within the framework of Chapter 2, we will exhibit, in an educational way, the basic models for GaN-type technologies. The extraction of the electro-optical model will provide a base that will be used in Chapters 3 and 4. We will more specifically develop the physical and chemical analyses relative to the polymer materials research. Indeed, this type of analyses being less common in microelectronics, it seemed sensible to emphasize the description of these parts. In addition, we will justify in Chapters 3 and 4 the great interest of these physicochemical analyses. The more conventional analyses will also be addressed in a more succinct manner.

In order to organize all of these manipulations, a methodology was established. The latter, allowing for the extraction of the parameters of the studied device, will constitute the architecture of the physical failure analyses developed in Chapters 3 and 4.

2.1. Junction temperature measurement methodologies

The junction temperature, JT, of an optoelectronic component represents the temperature of a component's active area. This parameter is to be distinguished from the temperature called "package temperature", T_P, which represents the temperature of the encapsulating housing.

For components of low dissipated power (<100 mW) and reasonable thermal resistance (<200 K/W), we can consider that $T_J \approx T_P$.

For components of high-dissipated power (>100 mW), the temperature difference $T_J - T_P$ is no longer negligible. In this case, we have self-heating of the chip. Since the functional and physical parameters depend on T_J and that only T_P is known during the manipulation, it is important to develop methods to determine T_J.

A method of determining T_J using the model of spontaneous emission and the optical power was proposed by Béchou et al. [BEC 08]. In this research, two methods for determining T_J have been developed:

– an electrical method developed in section 2.1.1;

– a detailed optical method detailed in section 2.1.2.

These two methods are complementary and allows us to ensure the validity of measurements. The temperature difference within an assembled component can be modeled by a thermal circuit using the electrical/thermal analogy. The latter is presented in Figure 2.1.

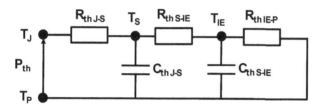

Figure 2.1. *Circuit diagram equivalent to the thermal model of an LED*

$R_{th\ JS}$, thermal resistance between the junction of the chip and its support; $R_{th\ S\text{-}IE}$, thermal resistance between the support of the chip and the electrical insulator; $R_{th\ IE\text{-}P}$, thermal resistance between the electrical insulator and the ambient air; $C_{th\ JS}$, thermal capacity between the junction of the chip and its support and $C_{th\ S\text{-}IE}$, thermal capacity between the chip's support and electrical insulation.

By placing themselves in a steady state, the thermal capacities are negligible. From this equivalent circuit diagram, the total power that is dissipated P_{dis} is deduced from equation [2.1] as follows:

$$P_{dis} = \frac{T_J - T_P}{R_{th}} \qquad [2.1]$$

R_{th} is the total thermal resistance (sum of three thermal resistances $R_{th\ J\text{-}S}$, $R_{thS\text{-}IE}$ and $R_{thIE\text{-}P}$). The dissipated power can also be estimated by taking into account that the temperature rise caused by the vibration of the network when the LED is powered (phonons → electric power) differs from light emission. This relationship is described by equation [2.2] as follows:

$$P_{dis} = P_{electrical} - P_{optical} = V_d I_d - P_{optical} = V_{th} I_d + R_S I_d^2 - P_{optical} \qquad [2.2]$$

$P_{optical}$ is the optical output power of the encapsulated LED, $P_{electric}$ is the electric power of the LED, RS is the series resistance, V_d is the voltage at the LED terminals, V_{th} is the threshold voltage and I_d is the LED power current.

2.1.1. Electrical methods

2.1.1.1. Bench measurements

The principle is to measure the change in the junction temperature based on the current going through the temperature-regulated LED. The measurement of the junction temperature allows us to highlight the chip's self-heating phenomenon. This phenomenon is caused due to the efficiency of a light-emitting component. The first simple definition of efficiency is the ratio between the number of photons collected and the number of injected electrons. The value of the efficiency being approximately 10%, a large part of the energy injected into the LED is dissipated by the Joule effect. It has been shown that the voltage across an LED was dependent on the junction temperature [SCH 06, HEL 08]. Here, we are interested in voltages above the diode's threshold voltage. By observing the variation in this voltage across the LED according to the current, we can determine the junction temperature. Finally, it is important to know the case temperature (T_P). The LED will be placed in a liquid nitrogen cryostat to realize this last operation. The benchmark block diagram is shown in Figure 2.2.

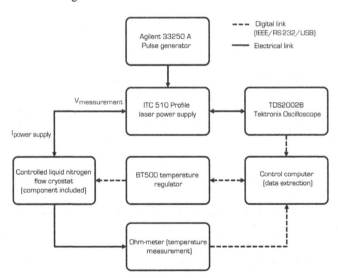

Figure 2.2. Block diagram of the measurement bench of an LED's junction temperature

The bench consists of the following devices:

– a PROFILE ITC 1 510 A/6 V MAX laser power with a bandwidth of 200 kHz for a current resolution of 10 uA. The latter is modulated by a pulse generator to prevent LED overheating;

– an Agilent 33250A pulse generator delivering voltage pulses to the laser power. Its range in pulse width is 8 ns to 1,999.9 s for a period of 20 ns to 2,000 s. The transition time ranges from 5 ns to 1 ms with overshoot lower than 5%;

– a digital TEKTRONIX TDS 2002B oscilloscope for viewing the sent pulses and to recover the voltage variations. The bandwidth is 60 MHz with a sampling rate of 1 GS/s. The accuracy of the voltage rating is less than 3% with a rise time less than 5.8 ns;

– a LN$_2$ flow cryostat is controlled wherein the component is located;

– a temperature control unit (BT 500) used for regulating the temperature T$_P$ (package temperature);

– an ohm meter giving the value of the resistance of the PT100 probe in order to know the T$_P$ of the LED.

To overcome the parasitic resistance induced by the electrical cables, the LED is connected in measure with four wires with Triax cables.

2.1.1.2. Principles of the electrical method and the associated models

This measurement method is based on the method of voltages [HON 04, GU 04]. The temperature regulation can be applied only on the casing; the principle of this measurement is based on the self-heating mechanism.

The first step of the methodology consists of measuring the LED terminal voltage's variation at the temperature (T$_P$) while avoiding temperature variations due to the chip's self-heating. It is thus under the following condition: T$_J$ ≈ T$_P$. The LED is continuously supplied with a sufficiently low current (≈1 mA) to neglect this self-heating. This allows determining the K coefficient (V/K) from the characteristic of Figure 2.3.

Linear regression applied to this graph can yield equation [2.3] [BEC 08, HON 04]:

$$V_{I_{cc}} = -K.T_p + V_0 \qquad [2.3]$$

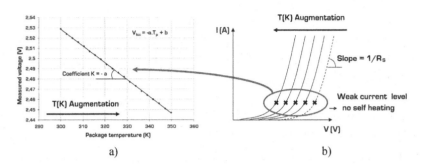

Figure 2.3. *a) Characteristic $V_{Icc} = f(T_P)$: determination of the coefficient K; b) diagram of the curve I(V) for determining the voltage V_{Icc}*

The coefficient K is therefore extracted from this first empirical relationship. Based on studies conducted by Hong *et al.* [HON 04, GU 04], junction temperature, T_J, of an LED is calculated from the voltage variation in temperature at a reference condition. By hypothesizing that self-heating is negligible when the LED is powered at 1 mA (Figure 2.3b), the expression of T_J is given by equation [2.4]:

$$T_j = T_{REF} + \frac{V_{test} - V_{REF}}{K}$$

[2.4]

T_{REF} is the reference ambient temperature, V_{REF} is the reference voltage at T_{REF} and V_{test} is the measured voltage in pulses (corresponding to I_{test}).

The second step is therefore based on the estimation of T_J (equation [2.4]). To perform this estimation, we measure the voltage V_{test}, which accounts for the increase in T_J as a result of the supply current I_{test}. The measurement is therefore carried out in pulsed mode to neglect the LED's cooling. The thermal capacities are also negligible since the applied pulse is too short compared with the thermal relaxation time of the active zone's materials.

The pulse is thus negative and is defined according to the following reference conditions:

– an ambient reference temperature (T_{REF}) constant throughout the duration of the measurement;

– the different corresponding power currents (I_{test});

– a reference current (I_{REF}) low enough to neglect self-heating. This current corresponds to the voltage V_{test} from which T_J is determined.

Figure 2.4 shows the diagrams of the pulse used for this second step.

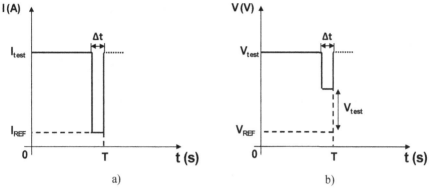

a) b)

Figure 2.4. *a) Modulation of the LED's current supply and b) impulse response of an LED (measurement of V_{test})*

For each power current I_{test}, we deduce the voltage V_{test} (Figure 2.4b), which allows getting the final characteristic of Figure 2.5.

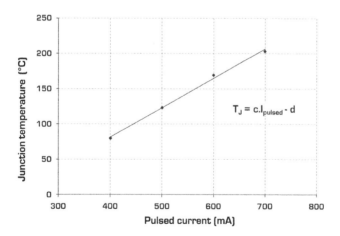

Figure 2.5. *Final characteristic $T_J = f(I_{pulsed})$*

The determination of T_J by this method allows characterizing the component at the desired T_J. The dimension of the pulse width Δt is a step that must be adapted to each type of component, so that the LED's current decreases until it reaches IREF.

The transition frequency depends, first, on the active area's volume. The current response of an LED will be highly dependent on this volume. This setting is made in manipulation and the typical pulse width is 5 μs for a frequency of 200 Hz.

2.1.2. Optical methods

The principle of determining the junction temperature with the optical emission spectrum is based on the expression of the spectral distribution R_{spon} (hv) of the radiative recombination rate ltR_{spon}, i.e. equation [2.5] [ROS 02].

$$R_{spon}(h\nu) = K_{spon}(h\nu - E_g)^{1/2}\exp\left(-\frac{h\nu - E_g}{kT_J}\right)$$ [2.5]

K_{spon} is the spontaneous emission coefficient.

The interesting part is the exponential function, which is directly connected to the junction temperature. We show that, for high energies, the term E1/2 is negligible compared with the exponential term. Figure 2.6 shows the superimposition of the theoretical F (E) and experimental spectrum in a semi-logarithmic scale.

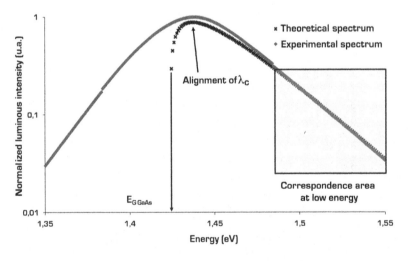

Figure 2.6. *Alignment of theoretical and experimental spectra: optical spectra in logarithmic scale of a GaAs LED @ T_J = 300 K at 100 mA*

We note that, for energies above 1.48 eV, the two curves are superimposed. Determining T_J follows naturally. It is also observed that the semi-logarithmic curve is a straight line. This reflects an exponential type of behavior and justifies the model presented by equation [2.5]. Note that the exponential is decisive when compared with the square root of F.

This method is valid for GaAs materials with parabolic bands. In the case of GaN, a non-parabolic band model is developed before applying this method. On the other hand, the light/encapsulation materials interaction does not allow using this model. We will develop this aspect in sections 2.2 and 2.3 of this chapter.

2.1.3. Methodology synthesis and thermal parameters

Two methods have been proposed in this section to determine the junction temperature of an optoelectronic component. The main objective of these two methodologies is identical: to characterize the studied component at a known T_J.

The electrical method consists of determining the junction temperature by impulse measurement. This method results in extraction of T_J, which also allows determining the casing temperature to be applied in order to realize measurements at the desired T_J.

The optical method is based on the theoretical model of the optical spectrum at a fixed T_J. The superposition of the theoretical spectrum and the experimental spectrum, on a semi-logarithmic scale and for high energies, validates the fact that the measurement is at the same T_J as that chosen in the theoretical model.

The optical method implies that we know the structure of the component's active area and its physical properties, whereas the electrical method requires no knowledge with respect to the component but may be limited by the accuracy of the measurement based on the devices used (e.g. oscilloscope). The measurement can be accurate to ±2 to 5 K depending on the type of oscilloscope. These two methods are complementary and produce the same result, as shown in the general flowchart of the methodology for measuring the junction temperature of an LED shown in Figure 2.7.

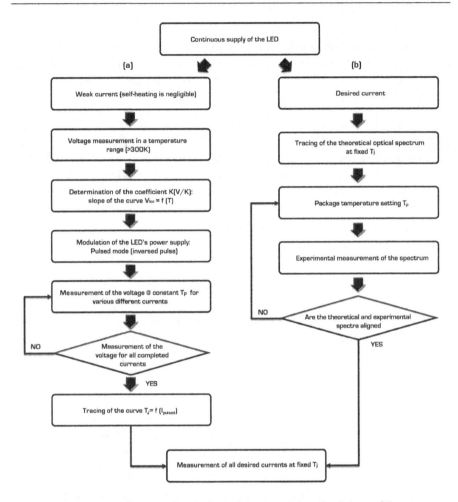

Figure 2.7. *Organizational chart of the general methodology of the measurement of T_J of an LED: a) electrical and b) optical methods*

Table 2.1 summarizes the important thermal parameters recalling their theoretical equations, their typical values and their location within the component.

Parameter	Theoretical expression	Localization in the component	Typical value of an AlGaAs/GaAs DH LED @ $I_d = 100$ mA and $T_p = 300$ K	References
T_J	$T_P + \dfrac{P_{dis}}{R_{th}}$		$315 - 340$ K	[DES 02, BEC 08, VEY 10]
$P_{electrical}$	$V_{th}I_d + R_S I_d^2$		$140 - 160$ mW	[DES 02, BEC 08, VEY 10]
V_{th}	$V_\varphi - R_S I_d$		$1.3 - 1.4$ V	[DES 02, BEC 08, VEY 10]
R_S	$\dfrac{V - V_{th}}{I_d}$		$1.7 - 2.4 \ \Omega$	[DES 02, BEC 08, VEY 10]
R_{th}	$\dfrac{T_J - T_P}{P_{dis}}$		$85 - 200$ K/W	[DES 02, BEC 08, VEY 10]
P_{dis}	$P_{electrical} - P_{optical}$		$120 - 140$ mW	[DES 02, BEC 08, VEY 10]
$P_{optical}$	$\eta_{ext} \dfrac{J}{q} h\nu$		$18 - 22$ mW	[DES 02, BEC 08, VEY 10]
T_P	-		300 K	[DES 02, BEC 08, VEY 10]

Table 2.1. *Important thermal parameters located on the different areas of an LED*

2.2. Mechanisms and electrical models of an LED

Electrical characterization is one of the main tools for nondestructive analysis of an encapsulated LED. It can identify and extract electrical parameters according to the chip's technology. These parameters can be linked to functional electrical parameters of the diode (current and threshold voltage). The junction temperature is determined by the evaluation of the junction-package thermal resistance (See section 2.1).

The current–voltage measurement (characteristic I(V)), from which electric models described in section 2.2.2 are built, allows to highlight electronic transport phenomena.

2.2.1. *Current–voltage measurement bench I(V)*

The basic principle is to measure the current's variation as a function of the voltage applied across an LED that is regulated in temperature. The bench block diagram is shown in Figure 2.8. The temperature, controlled by this bench, is the external temperature of the LED's assembly (T_P).

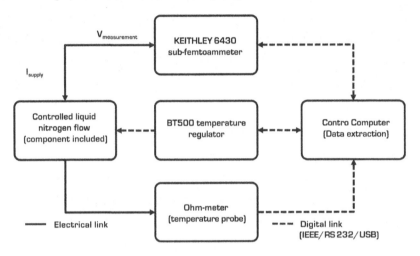

Figure 2.8. *Block diagram of the measurement bench I(V)*

The equipment used consists of:

– an analyzer with semiconductor parameters KEITHLEY 6430 connected by an IEEE bus connected to the CPU of the control computer. This device consists of a current source (10–16 A to 0.1 A) of resolution 10–17 (error 0.1%) and a voltage source (0–10 V) of resolution 10-6 V (error 0.1%);

– a controlled LN_2 flow cryostat wherein the component is located. It allows temperature regulation in a range of 80–350 K with a precision of 0.1 K;

– a temperature control unit (BT 500 temperature controller) used for temperature regulation during measurements. It controls the heating resistor of the cryostat using a PID (proportional integral derivative) automatic system;

– a dry (ADIXEN) pump whose role is to create a primary vacuum (10^{-2} Torr) in the vacuum chamber of the cryostat;

– an ohmmeter giving a resistance value denoted by R_{sensor}, corresponding to the value of the resistance of the PT100 heat sensor. This probe gives access to the LED package temperature.

To overcome the resistance of electrical cables, the LED is connected in measurement with four wires with Triax cables (Keithley) [SIN 01, GRU 06].

2.2.2. Electronic transport phenomena

Electric models used for GaN LEDs differ in certain aspects from those of GaAs LEDs because of the metal/semiconductor type of contact. This section complements the electronic transport phenomena applied to a GaN structure and clarifies the points that it differs in from GaAs LEDs.

The characteristic I(V) typical of an InGaN/GaN multi-quantum well LED (MQW LED) is shown in Figure 2.9.

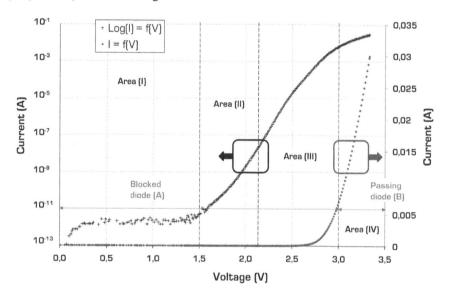

Figure 2.9. *Characteristic I(V) of an InGaN/GaN MQW LED*

The big difference between the I(V) of a GaAs LED and that of a GaN LED (Figure 2.9) comes from the difference of metal/semiconductor contact. For a GaAs and GaN LED, contacts are generally ohmic. However, for some GaN LEDs such as those studied in this book, the contact is not perfect and has a Schottky-type character. The latter gives rise to thermionic and tunnel currents that mask the classical recombination currents visible in the structure of GaAs LEDs.

The first plot I(V) (Figure 2.9) is classical, in a linear scale, and shows a threshold voltage and two main areas: one where the diode is conducting (V > Vth) and another where the diode is blocked (V < Vth). The behavior is one of a diode.

The second electrical characteristic in a semi-logarithmic scale distinguishes four current injection regimes showing, for low injection levels, nonlinearities:

– very low injection level → shunt (I): I ≤ 10 pA;

– low injection level → tunnel ET (II): 10 pA ≤ I ≤ 1 uA;

– medium injection level → thermionic ETT and ETE (III): 1 uA ≤ I ≤ 10 mA;

– high injection level → series (IV): I ≥ 10 mA.

Areas I, II and III represent 5% of the total optical power emitted by the LED, while the area IV alone represents 95%. We are thus interested in this section, to establish the analytical model of transport phenomena in two operating areas (II and III) of an MQW InGaN/GaN LED. Figure 2.10 shows the diagrams of the studied MQW structure.

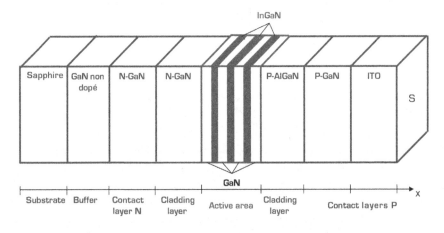

Figure 2.10. *Diagram of the studied InGaN/GaN MQW structure*

This MQW structure, typical of MQW InGaN/GaN LEDs with lateral emission on sapphire substrate (see Chapter 1), is that of studied GaN LEDs. In the case of this structure, only the low and medium current injection levels (10 pA ≤ I ≤ 10 mA) differ from those of the GaAs-type DH. Indeed, the layers close to the contacts create a Schottky character potential barrier [GRU 06]. The latter is home to three currents: the tunnel current, the tunnel current assisted by thermal effect and the thermionic current. These three currents are in series with the SRH and photonics currents. The tunnel current's equivalent resistance is dominant compared with the classic resistance from the SRH and photonics currents. We essentially observe these interface phenomena.

The barrier thus created is shown in Figure 2.11 and features a Schottky-type contact.

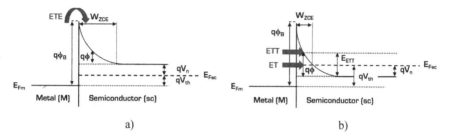

a)   b)

Figure 2.11. *Diagram of band of metal/semiconductor contacts creating a Schottky barrier: a) thermionic effect (ETE) and b) tunneling assisted thermal effect (ETT) and tunneling (ET)*

The majority carriers can then cross the potential barrier thanks to three different mechanisms [MOR 08b, SZE 07]:

– By thermionic emission (ETE) for weakly doped semiconductors ($N_A \leq$ 10^{17} cm^{-3}). The depletion area (ZCE) is relatively empty because its width, W_{ZCE}, is large. This implies that it is virtually impossible for holders to cross the barrier unless they are helped by intrinsic defects. However, holders may pass the barrier's height ($q\phi_B$), if the supply voltage (qV) is greater than the latter (Figure 2.11(a)).

– By tunnel effect assisted by a thermal effect (ETT) for moderate doping (10^{17} cm^{-3} $\leq N_A \leq 10^{18}$ cm^{-3}). Here, the width of the ZCE is not small enough for a tunnel effect and there must be sufficient energy for the holders to attain a level, where the ZCE is small enough (Figure 2.11(b)).

– By tunnel effect (ET) for semiconductors with a high doping level ($N_A \geq$ 10^{18} cm^{-3}). The ZCE is in this case sufficiently narrow so that the carriers can pass directly through it (Figure 2.11(b)).

The barrier height $q\phi_B$ can be calculated from the electronic properties of metal and semiconductor using equation [2.6].

$$q\varphi_B = q\chi_{SC} + E_{g\,SC} - q\varphi_{metal}$$ [2.6]

ϕ_{metal} is the barrier height of the metal layer, χ_{SC} is the electronic affinity of the semiconductor SC and Eg is the bandgap (or "gap") of the semiconductor.

The width of the WZCE space charge region can be expressed using equation [2.7].

$$W_{ZCE} = \sqrt{\frac{2\varepsilon_s}{qN_A}(\varphi_B - \varphi)} \qquad [2.7]$$

φ is the semiconductor barrier height, ε_s is the semiconductor dielectric constant and N_A is the concentration of acceptor atoms (holes). This expression indicates that the ETE, ETT and ET effects are activated depending on the variation in the width of the ZCE (Figure 2.11).

2.2.2.1. Thermionic effect: SUMMER

By hypothesizing that the ZCE is wide enough, the traditional expression of current density, related to ETE regime, is approximated by an exponential law described by equation [2.8]. This law is based on the condition that the assembly's series resistance (test bench) is negligible compared with that of the component, which is our case, since measurements are made in four wires with Triax cables:

$$J_{ETE} = J_{ETE\,0}\left[e^{\frac{qV_{th}}{kT}} - 1\right] \qquad [2.8]$$

$J_{ETE\,0}$ is the saturation value of the current density J_{ETE} being written in the form of equation [2.9]:

$$J_{ETE\,0} = A^* T^2 \exp\left(\frac{-q(\varphi_B - \Delta\ \varphi)}{kT}\right) \qquad [2.9]$$

A^* is Richardson's constant of the semiconductor and $\Delta\phi$ is the barrier height between the vacuum level and $q\phi_B$ described by equation [2.10]:

$$\Delta\varphi = \sqrt{\frac{qE}{4\pi\varepsilon_s}} \qquad [2.10]$$

E is the electric field at the metal/semiconductor interface given by equation [2.11]:

$$E = \sqrt{\frac{2qN_A}{\varepsilon_s}\left(-V_d + V_{bi} - \frac{kT}{q}\right)}$$ [2.11]

V_d is the diffusion voltage and V_{bi} is the LED supply voltage.

2.2.2.2. Thermally assisted tunneling: TAT

The ETT process exists throughout a range of temperatures that help carriers cross the potential barrier at energy E_{ETT} by the tunnel effect, as indicated in Figure 2.11(b). This electronic transport has been modeled by Padovani [PAD 66] and Stratton [STR 62] under the form of equation [2.12]:

$$J_{ETT} = \frac{A^*T^2}{2\pi kT}\left(\frac{\pi}{f_{ETT}}\right)^{\frac{1}{2}}\exp\left(\frac{qV_n}{kT} - b_{ETT} - c_{ETT}E_{ETT}\right)\left[1 + \text{erf}\left(E_{ETT}f_{ETT}^{\frac{1}{2}}\right)\right]$$ [2.12]

b_{ETT}, c_{ETT} and f_{ETT} are the Taylor expansion coefficients related to the transparency coefficient of the barrier around EETT. This expression has been rewritten by Morkoç [MOR 08b] in the form of equation [2.13]:

$$J_{ETT} = J_{ETT\,0}\left[e^{\frac{qV_d}{n_F kT}} - 1\right]$$ [2.13]

with $n_F = \frac{E_{00}}{kT}\coth\left(\frac{E_{00}}{kT}\right)$, $E_{00} = \frac{q\hbar}{2}\sqrt{\frac{N_A}{\varepsilon_S m_h^*}}$,

$0n_F$ is the ideality factor, E_{00} is the internal electric field, m_h^* is the effective mass of majority carriers, $\hbar$ is reduced Planck's constant and $J_{ETT\,0}$ is the J_{ETT} saturation value described by equation [2.14]:

$$J_{ETT\,0} = \frac{A^*T^2\sqrt{\pi qE_{00}(\varphi_B - V_d + V_n)}}{kT\cosh\left(\frac{E_{00}}{kT}\right)}\exp\left(\frac{qV_n}{kT} - \frac{q(\varphi_B + V_n)}{E_{00}\coth\left(\frac{E_{00}}{kT}\right)}\right)$$ [2.14]

2.2.2.3. Tunneling: ET

For low temperatures and a high doping level, the tunnel effect process dominates the current flow in the metal/semiconductor contacts. The current density by tunnel effect is presented by Padovani and Stratton in the form of equation [2.15]:

$$J_{ET} = A_{free}^* T^2 \left(\frac{E_{00}}{kT} \right)^2 \frac{\varphi_B - V_d}{\varphi_B} \exp \left(\frac{-2(q\varphi_B)^{3/2}}{3E_{00}\sqrt{q\varphi_B - qV_d}} \right) \qquad [2.15]$$

2.2.2.4. Electrical parameters of a GaN LED

Table 2.2 lists all the electrical parameters extracted during the analyses of the characteristic I(V) of a MQW InGaN/GaN LED. The measurements are controlled for temperature using the cryostat at a junction temperature of TJ = 300 K in order to construct a valid electrical model for a constant temperature TJ.

Parameter–working zone	Theoretical expression	Localization in the component	Typical value of an InGaN/GaN MQW LED @ I_d = 30 mA and T_p = 300 K	References
R_{Shunt} – I	$\dfrac{V - R_s I_d}{I_{shunt}} \approx \dfrac{V}{I}$		$4.10^{11} - 7.10^{11}\ \Omega$	[MOR 08c, MOR 08b, DES 10]
φ_B – II and III	$q\chi_{SC} + E_{g\,SC} - q\varphi_{metal}$		1.6 – 1.7 eV	[MOR 08c, MOR 08b, DES 10]
E_{00} – II and III	$\dfrac{q\hbar}{2}\sqrt{\dfrac{N_A}{\varepsilon_s m_h^*}}$		80 – 100 meV	
V_{th} – IV	$V_\varphi - R_s I_d$		1.82 – 1.83 V	[MOR 08c, MOR 08b, DES 10]
R_s – IV	$\dfrac{V - V_{th}}{I_d}$		8.3 – 8.6 Ω	

Table 2.2. Important electrical parameters located on the different areas of MQW LED

ZCA is the contact area N, ZA is the active area and ZCP is the contact area P.

The electrical parameters are attached to the different operational areas of the LED (Figure 2.9). These will be used to describe the electrical behavior of GaN LEDs, in Chapters 3 and 4, when undergoing aging.

2.3. Mechanisms and optical models of LED

The optical characteristics giving access to the main functional parameters are the optical power and the central wavelength (optical spectrum). They supplement the electrical analysis to build an electro-optical model used to understand the operation of an LED powered at $I > I_{th}$. In order to have an additional optical characteristic to the electrical characteristic, the measurement procedures must be suitable for controlling the junction temperature T_J. Indeed, in both cases, T_J must be identical to keep the parameters related to materials constant. T_J measurement techniques are used to ensure this type of manipulation.

In the same manner as in section 2.2, an optical model based on different analyses of different characteristics will be built.

2.3.1. Bench optical power measurements

The principle of this measurement is to evaluate the output optical power of the encapsulated LED as a function of the supply current. A block diagram of this bench is proposed in Figure 2.12.

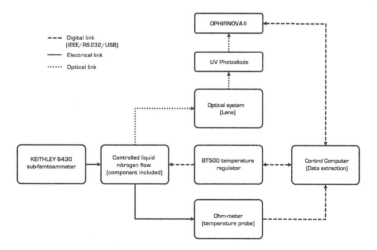

Figure 2.12. *Diagram of optical power measurements bench on the output of the cryostat*

The relative measurement of the optical power (at the cryostat's output) is useful for measuring optical powers at different junction temperatures, particularly at 300 K. It is composed of the following:

– a controlled LN_2 flow cryostat wherein the component is located;

– a temperature control unit (BT 500) used for controlling the room temperature;

– an ohm meter giving the value of the resistance of the PT100 to know the LED's temperature;

– AKEITHLEY 6430 semiconductor parameter analyzer for supplying the LED;

– an optical power measuring apparatus (NOVA OPHIR II). This system is provided with a UV photodiode covering a spectral range of 200 – 1,100 nm. Its measurement range is from 10 pW to 300 mW (error: ±6% 200 – 400 nm ±3% from 250 to 950 nm and ±5% from 950 to 1,100 nm). The latter is positioned in the alignment of the LED, in front of the quartz window of the liquid nitrogen cryostat. Data are transferred to the computer via RS232 link;

Figure 2.13 shows a typical measurement of the linear portion of the optical power of a GaAs LED at T_J = 300 K as a function of the supply current.

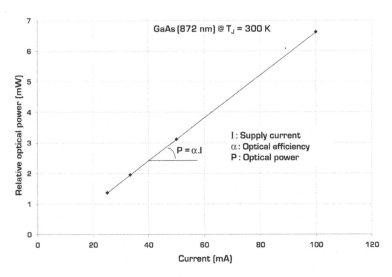

Figure 2.13. *Relative measurement of optical power of a GaAs DH LED @ T_J = 300 K*

The physical principles associated with characteristic P (I) will be mainly detailed.

2.3.2. Model of optical power

The optical power is the *main operating parameter* of an LED. Several units are also used in technical documentation of LEDs to indicate their optical power. Table 2.3 summarizes the different reference optical quantities in the literature [ROS 02, PIP 05].

Physical value	Energy unit	Visual unit
Power (Flux)	W	Lumen (Lm)
Luminosity	$W.m^{-2}$	Lux $(Lm.m^{-2})$
Luminous intensity	$W.sr^{-1}$	$Cd.m^{-2}$

Table 2.3. *Units used in optics* [1]

The optical power is, in our study, expressed in energy units that are optical Watts. This measurement unit is widely used in the industry due to the fact that it allows us to characterize all sources ranging from the UV to the IR. Units such as the lumen or candela are used, for devices intended to be referenced by the human eye (TV, public lighting, camera). These quantities are defined for wavelengths ranging from 380 to 760 nm corresponding to the visible radiation's spectrum. Measurement instruments used in the laboratory are calibrated in W. This is the reason why we will mainly use this unit for these measurements. On the other hand, the studied sources have wavelengths exceeding 760 nm (GaAs LEDs @ 872 nm).

The optical power is a quantity that depends not only on the light source but also on the path that the photons take between the source and the detector. We will describe, in this paragraph, the physical phenomena explaining the different losses of optical power. They will be classified into two major categories: reflection losses and absorption losses. We will mention luminescence yield or, more particularly, optical efficiency.

Figure 2.14 shows the diagram of an LED highlighting areas related to the optical power.

---

1 1 W = 683 Lm and 1 Lm = 1 $Cd.sr^{-1}$.

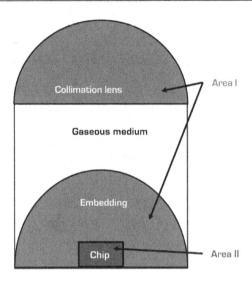

Figure 2.14. *Mapping areas of an encapsulated*
LED connected to the optical power

Zone I corresponds to the packaging located on the optical axis allowing light guidance. Area II sets the transmitter chip.

2.3.2.1. *Principle luminescent semiconductor materials*

The optical power is defined by the photon flux emitted by φ at the volume V of the chip's active area. This flow is described by equation [2.16].

$$\varphi = R_{spon} \frac{V}{S} = \frac{n}{t_{rad}} d \text{ with } R_{spon} = \int_{E_g}^{\infty} R_{spon}(E)dE \qquad [2.16]$$

S is the surface of the active area, d is the thickness of the active zone, t_{rad} is the radiative recombination time and R_{spon} is the total rate of radiative recombination.

When injecting carriers into an LED, not all holders provide luminescence. The active area is not perfect; the presence of deep defects leads to non-radiative losses for carriers that recombine on trap levels. This mechanism is directly dependent on the materials composing the active area, generally referred to as the material's optical efficiency. To quantify this phenomenon, it is necessary to introduce the internal quantum efficiency (or luminescence yield) η_i defining the ratio between the number of photons created by the cavity of volume V and the total number of injected carriers. It is expressed as equation [2.17] [ROS 02]:

$$\eta_i = \frac{t_{tot}}{t_{rad}} ; \frac{1}{t_{nrad}} = A_{nr} ; \frac{1}{t_{rad}} = Bn \qquad [2.17]$$

t_{tot} is the total recombination time, t_{nrad} is the non-radiative recombination time and A_{nr} is the non-radiative recombination rate due to the presence of defects.

Given the above expression, the flow of photons emitted by the chip can then be written as follows (equation [2.18]):

$$\varphi = \eta_i \frac{J}{q} \qquad [2.18]$$

The internal quantum yield shows the carriers flow transformation (J/q) in photon flow φ. The order of magnitude of η_i for a AlGaAs/GaAs DH LED emitting at 870 nm is approximately 50% [ROS 02].

2.3.2.2. Losses by reflection

The emission of an LED is not isotropic but rather of Lambertian type; all the light emitted by the active area is not recovered outside of the semiconductor. Figure 2.15 shows the various possible reflections for photons emitted from the active area of a GaAs DH LED.

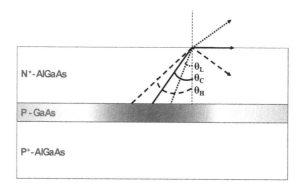

Figure 2.15. *Emission principle of an AlGaAs/GaAs DH LED with the various possible reflections on the surface of the active area. θ_C, critical angle; θ_L, low angles (<. θ_C) and θ_H, large angles (> θ_C)*

Several physical mechanisms may be involved and contribute to the optical power losses by reflection [ROS 02]:

– for angles less than the critical angle ($\theta < \theta_c$), the incident rays pass through the active area to DH by refraction. The fact remains that the low-angle rays undergo reflection at the active/coating area interface. We can define the reflection yielded using conventional laws (Fresnel's law and Snell's Law);

– for angles greater than the critical angle ($\theta > \theta_c$), the incident rays undergo total reflection. If a second phenomenon does not come into play, the yield is only 4%. In fact, part of the reflected photons are absorbed in the electron-hole pairs and re-emitted.

2.3.2.3. Losses by absorption

Optical absorption corresponds to the part of light interacting with matter. This phenomenon results in a loss of optical power at the sample's output (Figure 2.16).

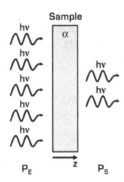

Figure 2.16. Principle of the absorption in a sample

The relationship for connecting the incident power P_E to the output power P_S is given by equation [2.19] [ROS 02]:

$$P_S = P_E \exp(-\alpha z) \qquad [2.19]$$

α: The sample's absorption coefficient.

The unit of α is expressed in cm^{-1} since the end of the exponential (here αz) is dimensionless. It thus reflects an evanescent wave because it is the square of the amplitude of the wave that decreases ($P = a.I^2$ where I is the wave's intensity). The wave's intensity decreases with the penetration depth of an absorbent material. The absorption coefficient depends on the energy of the incident photon and is written as a function of the gap of the material in the form of equation [2.20]:

$$\alpha(h\nu) = K_{abs}\left(h\nu - E_g\right)^{\frac{1}{2}}$$ [2.20]

K_{abs} is the absorbed photon energy constant [ROS 02].

This appearance indicates that the materials can be transparent at certain wavelengths (hv<Eg). This implies design rules of an optoelectronic device and component materials in order to minimize losses absorption [KIM 05a, RES 05]. Table 2.4 summarizes typical absorption coefficient values for three different wavelengths.

Target material	Gap energy (eV)	Incident wavelength (nm)	Absorption coefficient (cm⁻¹)	Optical index	References
GaAs	1.424	870	10^3	3.3	[CAS 75]
GaP	2.26	650	10^4	3.02	[PIM 11]
GaN	3.41	365	$<10^3$	2.3	[MOR 08c]
Sapphire	8.1 – 8.6	465	8	1.75	[MOR 08a, CAO 04]

Table 2.4. *Typical values of the key parameters related to the optical absorption of three materials used in LEDs*

According to the emission wavelength and target material, absorption losses can be important. They can be seen in two main areas:

– at the transmitter chip level (area II in Figure 2.14) where they are minimized as the active area is often composed of a DH or a MQW structure. However, certain substrates such as GaAs may absorb some of the incident light (transmittance < 20% @ 0.9 microns) [SCH 06];

– at the encapsulation level (area I in Figure 2.14):

- the chip's coating is required to improve the reflection and minimize total internal reflection phenomena. The type of coating and its design around the chip are therefore fundamental to the optical power [ALL 08];

- the collimating lens, according to its chemical composition, may be a factor of absorption losses. For a 5 mm LED lens with plastic (epoxy), its transmittance ranges from 85% to 90% at 300 K in the visible range [SIL 07].

These losses are taken into account in the transmission yield detailed in the following section.

2.3.2.4. Total optical power

Figure 2.17 shows the diagram of a DH LED with a TO47 casing, illustrating the different reflection mechanisms involved in the encapsulating casing.

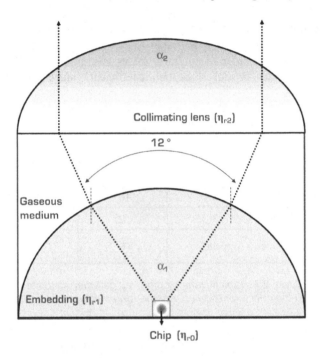

Figure 2.17. *Schematic optical losses within an encapsulated DH LED (case TO47)*

We thus define a transmission yield (η_t), which reflects the mechanisms of reflection losses (equation [2.21]):

$$\eta_t = \eta_{r0}.\eta_{r1}.\eta_{r2} \tag{2.21}$$

The η_t and η_i yields participate in defining a total external yield η_{ext}. It is expressed as equation [2.22]:

$$\eta_{ext} = \eta_t.\eta_i \tag{2.22}$$

This latter parameter allows connecting an LED's external optical power to the injected current's density J (equation [2.23]):

$$P_{ext} = \eta_{ext} \frac{J}{q} h\nu \qquad\qquad [2.23]$$

2.3.3. Bench spectral measurements

The principle of this measurement is to estimate the light intensity emitted as a function of energy (wavelength) at constant current and temperature. Figure 2.18 shows the block diagram of the analysis bench of optical spectra. The whole is mounted on a granite optical table on buffers, thereby isolating the mounting from external mechanical vibrations.

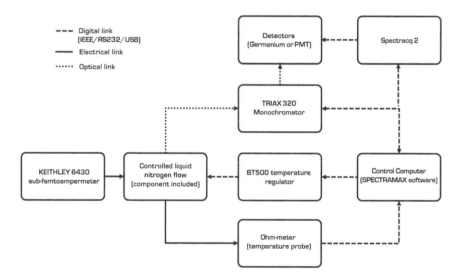

Figure 2.18. *System spectral measurements: TRIAX 320 monochromator*

The optical spectra analysis system comprises:

– a TRIAX 320 monochromator comprising two inputs and two outputs (axial and lateral), an input mirror, an output mirror and two networks covering a range of wavelengths from 300 to 1,800 nm:

- network 1: 1,200 lines/mm covering the range of 300–1,500 nm with a resolution of 0.26 Å;

- network 2: 900 lines/mm covering the range of 500–1,800 nm with a resolution of 0.35 Å.

The TRIAX 320 is a slit monochromator (inputs and outputs), with automatic opening from 10 microns to 2 mm, controlled by RS232 link. The focal length is 320 mm for a spectral dispersion of 2.35 nm/mm;

– a SPECTRACQ2 measurement acquisition system for converting data for a computer analysis by serial link between the sensors and the computer;

– a PMT (photomultiplier, in axial output, supplied with power) detector covering the range of 300–850 nm whose supply is controlled by the SPECTRACQ2 and connected to the latter by a coaxial cable in order to transfer the data. This sensor is used for GaN (blue and white) and GaP diodes (red). Its detection sensitivity is $1,43.10^{15}$ Jones (Jones $1 = 1$ cm.Hz^{-1}/2.w^{-1}) for a noise equivalent power (NEP) of 7.10^{-16} W.Hz^{-1}/2 at 400 nm;

– a germanium detector (output side, supplied with current) cooled with liquid nitrogen and used for GaAs diodes (IR). This one is controlled in the same way as the PMT with a range of 800–1,750 nm. Its detective sensitivity is 2.10^{13} Jones for a NEP of 5.10^{-14} W.Hz^{-1}/2 at 1,700 nm;

– a controlled LN$_2$ flow cryostat where the component is located;

– a (BT 500) temperature control unit used for room temperature Tp control;

– an ohm meter giving the value of the PT100 probe's resistance in order to know the Tp of the LED.

The measured light intensity not being absolute, spectral measurements are standardized. The bandwidth of the optical spectrum represents the part of the spectrum containing substantially all of the optical power. It is defined by the surface of the mid-height width $\Delta\lambda$. Figure 2.19 shows a typical standardized measurement L (λ) of a InGaN/GaN MQW LED powered at 30 mA at T$_J$ = 300 K.

Note that the typical optical spectrum of a GaN LED does not have a Gaussian form as pure as that of the GaAs. This phenomenon will be highlighted in Chapter 3.

Models of the optical spectrum are mainly constructed in energy; the optical spectrum will be represented as a function of energy rather than wavelength.

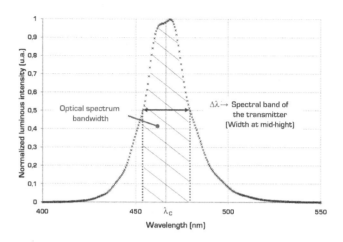

Figure 2.19. *Spectral measurement at T_J = 300 K L*
(λ) of a GaN MQW LED powered at 30 mA (I_{rated})

2.3.4. Phenomena of electronic transitions of a DH LED

This section attempts to synthesize the well-known optical spectrum's model of a DH LED resulting from electronic transitions [ROS 02, DES 02]. This optical characteristic differs depending on the type of transmitter chip's structure.

Figure 2.20 shows the three main phenomena of electronic transitions that usually compose an optical spectrum:

– the spontaneous emission;

– the optical cavity gain (stimulated emission);

– the Stark effect.

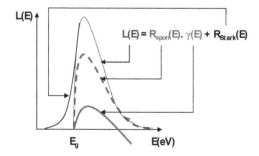

Figure 2.20. *Schematic of the theoretical spectrum of DH LED. For a color version of the figure, see www.iste.co.uk/deshayes/reliability2.zip*

The theoretical spectrum L (E), shown schematically in Figure 2.20, is modeled by the expression given by equation [2.24] [ROS 02].

$$\begin{cases} L(E) = R_{spon}(E).\gamma(E) & \text{for } E \geq E_g \\ L(E) = R_{stark}(E). & \text{for } E < E_g \end{cases} \qquad [2.24]$$

The following sections will present the models associated with spontaneous emission, the optical gain and the Stark effect. These three elements are used to model the full spectrum of a DH LED.

2.3.4.1. The spontaneous emission

Figure 2.21 presents a DH of a GaAs LED in real space in a strong injection level.

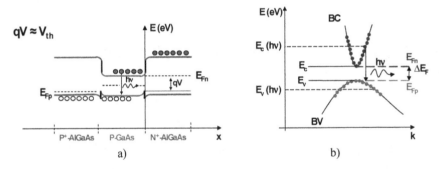

Figure 2.21. *Band diagram of a DH LED: a) representation of the active area in real space and b) representation of the active area in reciprocal space (k-space). For a color version of the figure, see www.iste.co.uk/deshayes/reliability2.zip*

When the high level of current injection regime is reached (V > Vth), a well of potential is created allowing to confine the carriers and thus to foster the radiative recombination (spontaneous emission). The representation E (x) determines the distribution of carriers in the structure (Figure 2.21a), while the representation in reciprocal space E (k) (Figure 2.21b) models R_{spon} (E).

By approximating Fermi–Dirac $f_c(E_c(hv))$ and $1 - f_v(E_v(hv))$ functions by Boltzmann functions, we can establish the model of spontaneous emission. The latter is calculated in energy (Figure 2.21(b)) from the expression of the spectral distribution of radiative recombination rate in a semiconductor given by equation [2.25] [ROS 02]:

$$R_{spon}(h\nu) = \frac{1}{\tau_R}\rho_j(h\nu)f_C(E_C(h\nu))(1-f_V(E_V(h\nu)))$$ [2.25]

τ_R is the radiative lifetime in the semiconductor given by equation [2.26]:

$$\frac{1}{\tau_R} = \frac{q^2\chi_{vc}^2 n_{op}\omega_{vc}^3}{\pi\hbar c^3\varepsilon_0}$$ [2.26]

ω_{vc} is the angular frequency of the incident photons, n_{op} is the optical index of the active area's material, c is the speed of light and χ_{vc} is the dipole matrix element that takes into account the fact that only the bands of heavy holes and light holes participate in an optical transition [DAK 06, BRE 99]. ρ_j (hv) is the density of electron-hole joint states relating to Einstein's equations [ROS 02].

The spectral distribution R_{spon}(hv) in energy is simplified considerably and results in the expression of equation [2.27] [ROS 02]:

$$R_{spon}(h\nu) = K_{spon}(h\nu - E_g)^{1/2}\exp\left(-\frac{h\nu - E_g}{kT}\right)$$ [2.27]

K_{spon} is the constant given by equation [2.28]:

$$K_{spon} = \frac{1}{2\pi^2}\left(\frac{2m_r}{\hbar^2}\right)^{3/2}\frac{1}{\tau_R}\exp\left(\frac{\Delta E_F - E_g}{kT}\right)$$ [2.28]

By integrating over hv the spectral distribution R_{spon}(hv) in energy, we obtain the expression of the total radiative recombination rate R_{spon} described by equation [2.29] [ROS 02].

$$R_{spon} = \frac{1}{\tau_R}\frac{N_J}{N_C N_V}np = Bn^2 = \frac{n}{t_{rad}} \quad :n = p \text{ (condition of the DH)}$$ [2.29]

2.3.4.2. *The optical gain*

Figure 2.22 shows the band diagram of the AlGaAs/GaAs DH in reciprocal space E (k) as well as the evolution of the absorption curves and the optical gain as a function of the position of the Fermi levels [ROS 02].

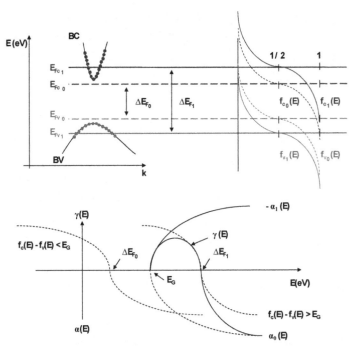

Figure 2.22. *Curves of evolution of the optical absorption α (E) and the optical gain γ(E) depending on the position of Fermi levels E_{FC} and E_{FV}. For a color version of the figure, see www.iste.co.uk/deshayes/reliability2.zip*

When the strong injection level regime is reached, part of the photons can meet the Bernard–Duraffourg condition, meaning that only the photons of energy less than ΔEF will be amplified. This means that the quasi-Fermi levels of electrons E_{Fn} and holes E_{Fp} are at greater energies than those of the gap and are present in the conduction and valence bands of the active area, respectively (Figure 2.22). This phenomenon corresponds to the optical amplification (or gain) and is called stimulated emission. Fermi Functions f_c(hv) and f_v(hv) delimit the area where absorption is negative, in other words where the optical gain γ becomes positive (for energies between E_g and ΔEF). It is given by equation [2.30]:

$$\gamma(h\nu) = K_{abs}\left(h\nu - E_g\right)^{1/2}\left(f_c(h\nu) - f_v(h\nu)\right) \qquad [2.30]$$

2.3.4.3. The Stark effect

The theoretical spectrum is usually superimposed on the real spectrum where we see that the part of energies less than E_g is not modeled. The latter refers to another effect named: the Stark effect [ROS 02, SZE 07, DAK 06]. Figure 2.23 illustrates the band diagram of the AlGaAs/GaAs DH LED during the Stark effect.

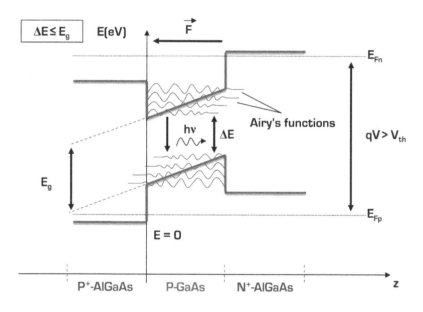

Figure 2.23. *Band diagram of an AlGaAs/GaAs DH LED with the Stark effect. For a color version of the figure, see www.iste.co.uk/deshayes/reliability2.zip*

The latter is due to the application of an electric field in the DH which modifies the slope of the valence and conduction bands, as well as the shape of the wave functions of electrons and holes. This phenomenon involves a non-zero probability of combining tunneling with radiative recombinations below the gap ($\Delta E \leq E_g$). The wave functions of the carriers are expressed by the solutions of equation [2.31]:

$$\frac{d^2 Ai(x)}{dx^2} - xAi(x) = 0 \qquad [2.31]$$

Ai (x) is the Airy's function and x is the direction perpendicular to the epitaxy plane.

For radiative recombination, Airy's function is exponential and is written in the form given by equation [2.32].

$$Ai(x) \approx \frac{1}{2\sqrt{\pi}x^{\frac{1}{4}}} \exp\left(-\frac{3}{2}x^{\frac{3}{2}}\right) \qquad [2.32]$$

The spontaneous emission on the Stark effect is then described by equation [2.33]:

$$R_S(\xi) = r_0\sqrt{\beta}.\pi\left[\left(\frac{dAi(-\xi)}{dx}\right)^2 + \xi Ai^2(-\xi)\right] \qquad [2.33]$$

with $\beta = \left(\frac{2m_r}{\hbar^2 q^2 F^2}\right)^{-\frac{1}{3}}$ and $\xi = \frac{h\nu - E_g}{\beta}$

F is the internal electric field applied to the active area (P-GaAs) [THO 02].

2.3.5. Optical parameters of a DH LED

Table 2.5 summarizes the important optical parameters for LED failure analysis.

Certain parameters summarized in this table are found in electric models. This demonstrates the complementarities of electro-optical analyses and validates the value of these parameters (ΔEF, τR or B). Indeed, the optical parameters were determined for very different injection levels, but for the same T_J (300 K). For electrical measurements, the current is much lower than 1 mA, which is not the case for optical measurements, where the current becomes much greater than 1 mA. In both cases, the material remains the same and therefore the associated parameters must remain constant.

Parameter	Theoretical expression	Location in the component	Typical value of an AlGaAs/GaAs DH LED @ $I_d = 100$ mA and $T_p = 300$ K	References
P_{ext}	$\eta_{ext}\dfrac{J}{q}h\nu$		$20 - 25$ mW	
η_t	$1 - \dfrac{\left(n_{opt\,ZA} - n_{enr}\right)^2}{\left(n_{opt\,ZA} + n_{enr}\right)^2}$		$30 - 40\%$	[DES 02, BEC 08, VEY 10]
η_{ext}	$\eta_t\eta_i$		$20 - 50\%$	
A	-		$8 - 10^5$ cm^{-1}	
K_{Spon}	$\dfrac{1}{2\pi^2}\left(\dfrac{2m_r}{\hbar^2}\right)^{3/2}\dfrac{1}{\tau_R}\exp\left(\dfrac{\Delta E_F - E_g}{kT}\right)$		10^4 to 10^5 cm^{-1}.eV$^{-1/2}$	
τ_R	$\dfrac{2\pi\hbar^2 c^3\varepsilon_0 m_e}{q^2 n_{op}E_g E_p}$		10^{-10} to 10^{-9} s	
B	$\dfrac{1}{\tau_R}\dfrac{N_J}{N_C N_V}$		10^{-11} to 10^{-10} cm^{-3}.s^{-1}	
χ_{vc}	$\left(\dfrac{\pi\hbar c^3\varepsilon_0}{q^2\tau_R n_{op}\omega_{vc}^3}\right)^{1/2}$		Few Å	
ΔE_F	-		$1.4 - 1.5$ eV	[ROS 02, DES 02, DES 10]
γ_{max}	$K_{abs}\left(h\nu - E_g\right)^{1/2}\left(f_c\left(h\nu\right) - f_v\left(h\nu\right)\right)$		10^3 to 10^4 cm^{-1}	
η_i	$\dfrac{t_{tot}}{t_{rad}}$		$40 - 50\%$	
t_{rad}, t_{nrad}	-		10^{-10} to 10^{-9} s	
B	$\left(\dfrac{2m_r}{\hbar^2 q^2 F^2}\right)^{-1/3}$		$50 - 200$ meV	
r_0	-		10^3 to 10^5	
F	-		10^6 to 10^7 V/m	

Table 2.5. *Important optical parameters located on the different areas of the DH LED*

2.3.6. *Phenomena of electronic transitions of a MQW LED*

This section presents the additional electronic transition phenomena of a MQW LED. Models of the Stark effect described for a DH LED are similar to a MQW LED. The well's energy levels and the parameter values are changing because of the materials that are different and the MQW structure. The objective of this section is to present how the models of spontaneous emission and optical gain are changed when studying a MQW structure (Figure 2.10).

Section 2.2 of this chapter shows that the threshold current is indirectly proportional to the thickness of the active area. This thickness corresponds to the size of the well of potential formed by the DH. To reduce this threshold current, it is necessary to reduce the thickness of the wells of potential until the quantification of the active area. This means that the wells are so narrow that the movement of carriers is quantified. Figure 2.24 shows a simplified band diagram (without taking into account the electric field) of the MQW structure of a GaN LED and the three main phenomena of electronic transitions that compose its optical spectrum.

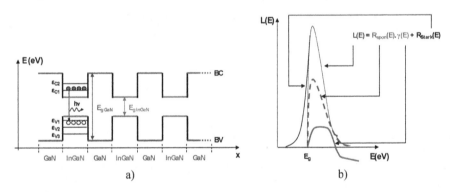

a)   b)

Figure 2.24. *a) Simplified band diagram of the InGaN/GaN MQW structure of a GaN LED and b) diagram representation of the theoretical optical spectrum of a GaN MQW LED. For a color version of the figure, see www.iste.co.uk/deshayes/reliability2.zip*

The conduction and valence bands of a MQW GaN LED are not parabolic. Considering this fact, the spectral distribution $R_{spon}(h\nu)$ is written in the form of equation [2.34] [ROS 02]:

$$R_{spon}(h\nu) = K_{spon}(h\nu - E_g)^{1/m} \exp\left(-\frac{h\nu - E_g}{kT}\right) \qquad [2.34]$$

where m is the coefficient reflecting the fact that the bands are not parabolic. This coefficient is also found in the expression of the spontaneous emission coefficient K_{spon} given by equation [2.35].

$$K_{spon} = \left(2m_r\right)^{1+\frac{1}{m}} \frac{1}{\pi \hbar \tau_R} \exp\left(\frac{\Delta E_F - E_g}{kT}\right) \qquad [2.35]$$

The expression of ΔE_F in a MQW structure differs from that of a DH. The carriers recombine on the energy levels of the quantum wells ($E_{C1,2}$ ... for the conduction band and $E_{V1,2}$ ... for the valence band). If the separation of the lower levels is much larger than the thermal energy kT ($E_{C2} - E_{C1} \gg kT$), and the total surface density of the carriers (ns) is not too large, a single sub-band (E_{C1}, E_{V1}) is populated, as shown in Figure 2.25 presenting the band and sub-band diagram of the studied MQW structure in real space, reciprocal space and the energy density of states [ROS 02].

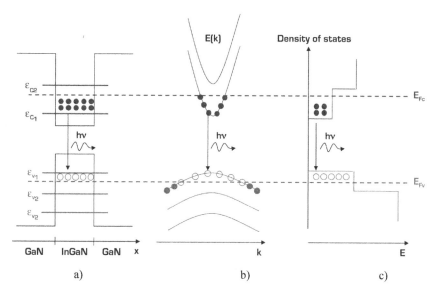

Figure 2.25. *Diagram of bands and sub-bands of the InGaN/GaN MQW structure of a GaN LED: a) real-space (x); b) reciprocal space (k); c) density energy of states (E). For a color version of the figure, see www.iste.co.uk/deshayes/reliability2.zip*

In this case, the Fermi levels can be expressed in the conduction band E_{FC} (equation [2.36]) and valence band E_{FV} (equation [2.37]).

$$E_{Fc} = E_{g_{InGaN}} + \varepsilon_C + kTln\left(exp\left(\frac{Jt_{tot}\pi\hbar^2}{qm_ckT} \right) - 1 \right) \qquad [2.36]$$

$$E_{Fv} = \varepsilon_V - kTln\left(exp\left(\frac{Jt_{tot}\pi\hbar^2}{qm_vkT} \right) - 1 \right) \qquad [2.37]$$

J is the current density of the LED.

As in the case of the DH, a part of the spontaneous emission is amplified by stimulated emission in the MQW. The optical gain, considered for an energy level (first sub-band) in the quantum well, is written in the form of equation [2.38]:

$$\gamma(h\nu) = \alpha_{2d}\left(f_C^1(h\nu) - f_V^1(h\nu) \right)\theta\left(h\nu - E_g - \varepsilon_C - |\varepsilon_V| \right) \qquad [2.38]$$

θ is Heaviside's function and f_C^1 is Fermi's distribution function in the conduction band given by equation [2.39]:

$$f_C^1(h\nu) = \frac{1}{1 + exp\left(\dfrac{E_C^1 - E_{Fc}}{kT} \right)} \qquad [2.39]$$

With $E_C^1(h\nu) = E_g + \varepsilon_{c_1} + \dfrac{m_r}{m_c}\left(h\nu - E_g - \varepsilon_{c_1} \right)$

f_V^1 is Fermi's distribution function in the valence band written in the form of equation [2.40]:

$$f_V^1(h\nu) = \frac{1}{1 + exp\left(\dfrac{E_V^1 - E_{Fv}}{kT} \right)} \qquad [2.40]$$

with $E_V^1(h\nu) = -\dfrac{m_r}{m_v}\left(h\nu - E_g - \varepsilon_{v_1} \right)$

α_{2d} is the absorption coefficient of the zero population quantum well defined by equation [2.41]:

$$\alpha_{2d} = \frac{2\pi q^2 \chi_{VC}^2 m_r}{\lambda_0 n_{sc} \varepsilon_0 \hbar^2 d}$$

[2.41]

λ_0 is the LED's wavelength in vacuum, n_{sc} is the refractive index of the semiconductor and d is the thickness of the quantum well.

2.3.7. Optical parameters of a MQW LED

Table 2.6 summarizes the important optical parameters for LED failure analysis.

Parameter	Theoretical Expression	Location in the component	Typical value of an InGaN/GaN DH LED @ I_d = 30 mA and T_p = 300 K	References
P_{ext}	$\eta_{ext} \dfrac{J}{q} h\nu$		2.5 – 2.7 mW	
η_t	$1 - \dfrac{\left(n_{opt\,ZA} - n_{enr}\right)^2}{\left(n_{opt\,ZA} + n_{enr}\right)^2}$		30 – 40%	[ROS 02, DES 02, DES 10]
η_{ext}	$\eta_t \eta_i$		20 – 50 %	
K_{Spon}	$(2m_r)^{1+\frac{1}{m}} \dfrac{1}{\pi\hbar\tau_R} \exp\left(\dfrac{\Delta E_F - E_g}{kT}\right)$		10^{-4} to 10^{-3} cm^{-1}.eV$^{-1/2}$	
τ_R	$\dfrac{2\pi\hbar^2 c^3 \varepsilon_0 m_e}{q^2 n_{op} E_g E_p}$		10^{-9} s	
γ_{max}	$\alpha_{2d}\left(f_C^l(h\nu) - f_V^l(h\nu)\right)\theta(h\nu - E_g - \varepsilon_C - |\varepsilon_V|)$		10^5 to 10^6 cm^{-1}	[ROS 02, DES 02, DES 10]
M	-		1.1 – 1.2	
ΔE_F	-		2.9 – 3.1 eV	
α_{2D}	$\dfrac{2\pi q^2 \chi_{VC}^2 m_r}{\lambda_0 n_{sc} \varepsilon_0 \hbar^2 d}$		10^{-5} to 5.10^{-5} m^{-1}	
ε_{C1}	-		150 meV	
ε_{V1}	-		– 40 meV	

χ_{VC}	$\left(\dfrac{\pi\hbar c^3 \varepsilon_0}{q^2 \tau_R n_{op} \omega_{vc}^3}\right)^{1/2}$		$0.4 - 0.5$ Å
B	$\left(\dfrac{2m_r}{\hbar^2 q^2 F^2}\right)^{-1/3}$		$60 - 180$ meV
r_0	-		$450 - 550$
F	-		10^7 at 10^8 V/m

Table 2.6. *Important optical parameters*
located on the different areas of the MQW LED

2.4. Physicochemical characterizations of an LED

The physicochemical characterizations allow, to a lesser extent, to help build the electro-optical models developed previously. Analyses of nuclear, electronic and optical types were used. These analyses were used to provide not only structural information but also information on the chemical compositions of materials of all components.

The chip's characterization helped develop an electro-optical model based, in large part, on the dimensions of the various layers and their epitaxial doping.

Regarding encapsulation, the physicochemical analyses helped clarify the thermal model and explain certain parasite luminescence phenomena such as silicone oil fluorescence.

The samples' preparation, critical phase of a physicochemical analysis, is divided into two parts:

– preparation of the bare chip;

– preparation of the encapsulation.

These analyses are often destructive and are therefore made in the initial stage to assist in the component's modeling.

This section exposes the methodology for performing sample preparation coupled with different physicochemical analyses. We briefly present the various analyses carried out in this project. Typical results will be presented to illustrate the contribution of each analysis in the study of an LED.

2.4.1. Sample preparation techniques

The sample preparation technique for a physicochemical analysis requires a relatively varied processes. Chemical etchings are used during the humid phase, mechanical ablation and collection of soft material. The sample preparation will be carried out according to the specifications of the physicochemical analysis. Thus, we will work, as appropriate, on the quality of the surface state, the part of a collection or the selectivity of a chemical etching to reveal the areas to observe.

2.4.1.1. Principle of a micro-section

This method involves carrying out a longitudinal cut of the encapsulated component in order to reveal its internal structure. Figure 2.26 schematically shows a micro-section of an encapsulated LED.

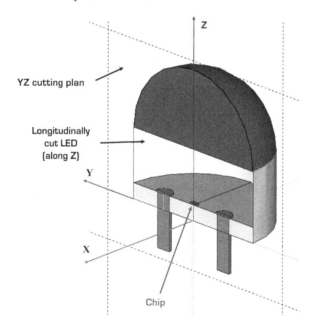

Figure 2.26. *Diagram of an LED cut along the Z-axis by micro-section*

The difficulty of such a manipulation lies in the determination of the cutting plane and the quality of the surface state. The summary diagram of Figure 2.27 summarizes the steps for carrying out a micro-section.

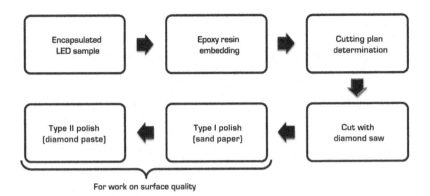

Figure 2.27. *Diagram of summary of steps for sample preparation by micro-section*

The sample is usually embedded in an epoxy resin mixed with a hardener and then vacuumed to remove present air bubbles. Once polymerized coating is done, diamond saw is then used in order to achieve a first longitudinal cut. The diamond is selected for its hardness (70 GPa) greater than all of the materials used in optoelectronics, for example, GaAs (7.56 GPa). We thereafter realize a polishing of the sample using paper silicon grains. This paper is in the form of 230 mm diameter discs with 500–2,400 silicon grains/disc for grain average diameters ranging from 9 to 30 μm. The polishing finishing is carried out with diamond grain-based dough of 0.25 μm average diameter. The grain size allows obtaining a surface quality suitable for scanning electron microscopy (resolution of a few nanometers).

2.4.1.2. Preparation of a "bare" LED

This preparation is designed to extract the chip from its encapsulation. The physicochemical analyses concentrated on the chip require this manipulation. The great difficulty is in removing the materials surrounding the chip while preserving it. Figure 2.28 shows the method of preparing an LED mounted on an OPTRANS T047 casing.

The first step consists of removing the LED's collimating lens in order to have access to the chip. This cutting is carried out with a cutter (thick diamond disc of 500 μm) attached to a high-speed rotary tool (15,000–35,000 r/min). The decapsulated LED is then placed in a 50 ml beaker, in which approximately 20 ml of potassium hydroxide-based selective solvent and 2-methoxyethanol (panasolve) are added. The silicone coating (polymer) is chemically etched humidly at a temperature of 400°C for several minutes until the boiling of the panasolve and the complete dissolution of the Si oil. After a first rinsing with deionized water, we dissolve the

(Au) bonding wires in aqua regia[2] for 10–15 minutes. A second rinsing with deionized water is applied before dissolving the silver-loaded adhesive joint (glue and silver mixture) with acetone for a few tens of seconds. This step is the most sensitive since it definitely separates the chip from its original metal support. With a pair of tweezers, we transfer (under an optical microscope) the chip from its original support to a glass dish filled with deionized water for the final rinsing. According to the physicochemical analysis, the chip is attached to a reference support (Si + Ag joint or carbon pellet).

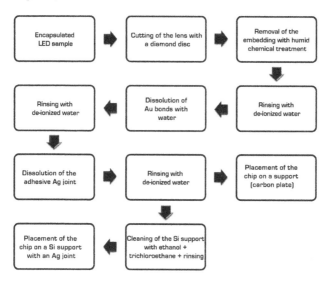

Figure 2.28. *Diagram of summary of steps for the preparation of a "bare" LED*

2.4.2. Nuclear analyses

Nuclear analyses are defined by the interaction of the incident particle (primary ions, α particles, UV photon) with the nuclei of atoms constituting the target material. In this section, the SIMS, RBS, ^{1}H NMR and MALDI-TOF mass spectrometry analyses will be presented.

2.4.2.1. Secondary ion mass spectroscopy

Secondary ion mass spectroscopy (SIMS) is a destructive analysis preferably adapted to inorganic material surfaces. It allows us to trace, by erosion of matter, the elements of the concentration profile in a semiconductor structure. We will therefore

---

2 Mixture of a dose of HNO_3 with three doses of HCl at the same concentrations.

measure the different thicknesses of the layers constituting a "bare chip" component, as well as, in some cases, the doping profile of the semiconductor's various layers.

An ion source emits a primary ion beam of several kiloelectron-volts with a current ranging from nanoamperes to several tens of microamperes according to the diameter of the sample to be analyzed and the type of ion used [VIC 89]. The primary ions are intended to excite the material and produce a de-excitation of the electronic states of the target material generally known as secondary electron emission. The primary ion beam is therefore chosen depending on the nature of the material to be analyzed. Traditionally, the following primary ions are used: Ar^+, Cs^+, O^{2+} or K^+. For example, an O^{2+} ion bombardment is more suited for a magnesium compound material (Mg) while the Cs^+ ions are more suited for a silicon-based material.

The SIMS (secondary ion mass spectroscopy) principle is illustrated by the diagram of Figure 2.29 [AGI 90].

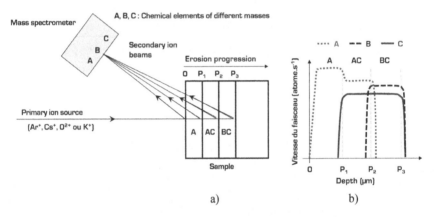

Figure 2.29. *a) Schematic diagram of the SIMS analysis and b) schematic of SIMS measurement*

The ion beam is shaped using magnetic optical (solenoids) to obtain spot diameters ranging in size from 50 μm to several millimeters. The energy deposited by the ion beam on the sample is such that a portion of the atoms are torn. These are found in the form of secondary ions that are extracted from the sample surface using an electric field created between the specimen and the extraction electrode. This new secondary ion beam is then focused to a mass filter system. In order to select the ions, we use a constant magnetic field B0. The trajectory of ions depending only on their mass makes it possible to make a selection of different atoms following their mass and thus the type of atom analyzed. This gives a mass spectrum of the sample's known elements. Thus, we can recognize the chemical elements that make up a semiconductor layer (case of LEDs). In some cases, we can go back to the doping of semiconductor layers. It is then necessary to have a reference spectrum of the doping element.

The SIMS analysis is also used in spatial analysis to determine the thickness of a sample's component layers with a resolution of 1–10 nm [WIL 89, BEN 87]. We show, in this case, quantum wells with a thickness ranging from 20 to 200 Å per well. This analysis is performed to confirm the validity of an already known structure. A measurement example is shown in Figure 2.30.

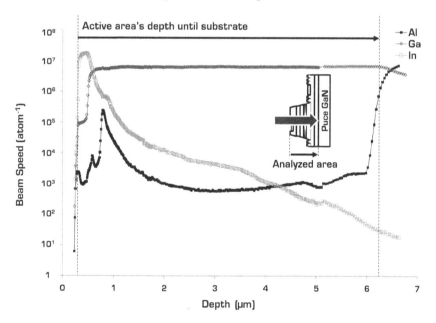

Figure 2.30. *Example of spatial measurement SIMS on a GaN LED*

This SIMS measurement has been obtained with O^{2+} ions for a current of 5 nA and a beam diameter of 50 μm on a depth of 6 μm. Table 2.7 summarizes the characteristics of a SIMS analysis.

Incident particle	Analyzed particle	Detected chemical species	Type	Resolution	Analysis depth
Primary ions $(Ar^+, Cs^+, O^{2+}$ or $K^+)$	Secondary ions formed by the interaction of primary ions /matter	All elements from boron	Qualitative, quantitative and destructive	10–100 Å	100 nm to few μm

Table 2.7. *Summary table of the SIMS analysis*

2.4.2.2. Rutherford scattering analysis

Rutherford scattering nuclear analysis RBS (Rutherford backscattering) is a nondestructive surface analysis used to identify a multilayer structure. Its block diagram is shown in Figure 2.31 with a measurement example [AGI 90, WAL 90, BRU 92].

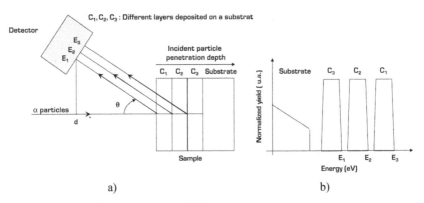

Figure 2.31. *a) Schematic diagram of an RBS measurement and b) schematic of an RBS measurement*

Incident particles α generated from light elements such as helium He^+ are sent to a target sample with an energy of 2–4 MeV. The elastic interaction with matter has the effect of reducing their energy according to the element and the penetration depth of the particle. RBS will be especially suited to the study of a multilayer system by giving its composition and thickness. The resolution in thickness is 10–50 nm.

The particles are backscattered, that is to say, their velocity vector is opposite to an angle θ. The principle is that of a billiard ball that hits the crystal lattice. We have a conservation of kinetic energy in this type of nuclear interaction since the incident energy is less than the energy of the nuclei bonds (a few GeV). The variation in the kinematic coefficient depends on the mass of the nucleus of the target atom and the detection angle. The particle detector is located at a distance "d" from the sample to capture the backscattered particles.

The main characteristics of the RBS analysis are summarized in Table 2.8.

Incident particle	Analyzed particle	Detected chemical species	Type	Resolution	Analysis depth
Particles α (He⁺)	Scattered particles (sample)	All elements starting from boron	Quantitative and nondestructive	$10 - 50$ nm	1 nm to 10 µm

Table 2.8. *Summary table of the RBS analysis*

2.4.2.3. *Proton nuclear magnetic resonance*

Proton nuclear magnetic resonance (NMR ^{1}H) is a destructive analysis (in our case) adapted to organic samples who, in this research, were extracted from the assembly of LEDs. This analysis uses the magnetic properties of hydrogen nuclei (protons) to provide information about the molecular structure of the sample by the resonance of nuclei under magnetic excitation [LAM 04, NAN 97, ERN 90]. The principle of this analysis is based on the nucleus resonance mechanism during the application of a magnetic field on the target sample. The atoms in the sample are identified by their resonant frequency to calculate their chemical shift δ in parts per million relative to a reference compound which is generally tetramethylsilane (TMS). The block diagram of the ^{1}H NMR is shown in Figure 2.32.

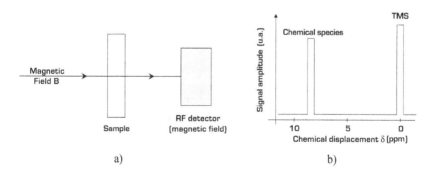

a)   b)

Figure 2.32. *a) Schematic diagram of the proton NMR and b) schematic of an NMR measurement*

This is a lower dosage quantitative method. It also has the advantage of providing dynamic information about the species in exchange. Its mass resolution is of a few parts per million. Two types of measurements are used in this book:

– the measurement of the chemical shift, given in Figure 2.33;

– the two-dimensional (2D) measurement of the diffusion coefficient (DOSY method for diffusion-ordered spectroscopy), shown in Figure 2.34.

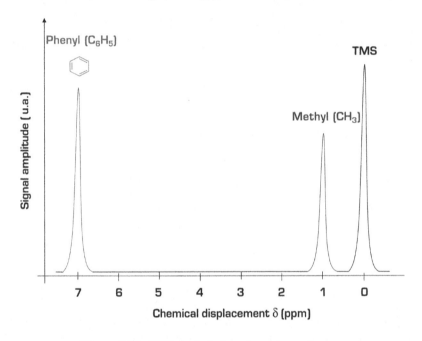

Figure 2.33. *NMR methyl phenyl spectrum diagram*

Each ray corresponds to the chemical shift of a molecule. In the example in Figure 2.33, a phenyl group is observed at 7 ppm and a methyl group at 1 ppm. Chemical shift tables identify the groups concerned.

The DOSY method is a method in two dimensions which allows discriminating components of a sample according to their diffusion coefficient (m^2/s). The diffusion axis becomes an additional spectroscopic axis to the chemical shift. Figure 2.34 shows a diagram of methyl and phenyl DOSY.

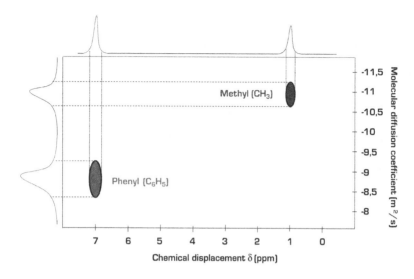

Figure 2.34. *Schematic of an NMR spectrum of methyl and phenyl DOSY. For a color version of the figure, see www.iste.co.uk/deshayes/reliability2.zip*

The surfaces shown in Figure 2.34 help differentiate the various molecules present in a sample by their movement and distribution. The characteristics of the ^{1}H NMR are summarized in Table 2.9.

Incident particle	Analyzed particle	Detected chemical species	Type	Resolution	Analysis depth
-	Proton (^{1}H)	Organic molecules	Qualitative, quantitative and destructive	Few ppm	Sample totality

Table 2.9. *Summary table of ^{1}H NMR*

2.4.2.4. Mass spectrometry (MALDI-TOF)

Mass spectrometry can be realized by various techniques. The destructive technique, part of this research, is matrix-assisted laser desorption ionization with a time-of-flight (MALDI-TOF) spectrometer.

Figure 2.35 schematically shows the principle of a MALDI-TOF spectrometer with an example of measuring an organic molecule [DE 07, MOH 04, COL 09].

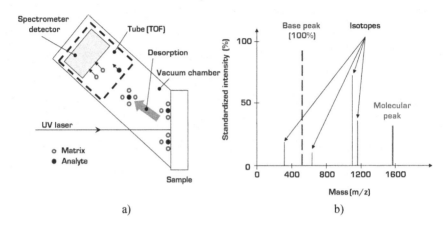

a)   b)

Figure 2.35. *a) Schematic diagram of a MALDI-TOF spectrometer and b) schematic of a mass spectrum. For a color version of the figure, see www.iste.co.uk/deshayes/reliability2.zip*

The MALDI-TOF spectrometry is divided into two phases:

– The sample (analyte) is dissolved in a solvent composed of small organic molecules that define the matrix. Such molecules must have a strong absorption around the wavelength of the used laser. In our case, it is a UV laser with a wavelength λ_C= 336 nm. The matrix-solution[3] is evaporated and then crystallized on a target sample holder. The result is a solid in which each analyte molecule is isolated. The analyte and the matrix are so-called co-crystallized.

– The second step is carried out under vacuum. A laser beam is irradiating the target sample and leads to the ionization of the analyte molecules: the absorption of photons by the matrix molecules leads to excitation of the electronic states of these. The energy relaxation in the solid leads to the ejection of matter (lift off) that gradually decomposes in the vacuum. The latter consists of several types of particles either neutral or ionized. The ions formed by the irradiation are then analyzed by the analyzer part (tube) of the time of flight spectrometer. This last part consists of measuring the flight time of ionized molecules by making them do a round trip in the tube. The molecules have a different flight time according to their mass. This flight time allows identifying the molecules constituting the original sample.

MALDI-TOF technique has the advantage of not needing large amounts of matter given its precision (a few parts per million) [DAS 07, EKM 09]. In the case of polymer study, it allows the determination of the chain ends (for a power

---

3 Mixing the matrix with the analyte.

resolution by mass of less than 20,000), the confirmation of the monomer unit, the calculation of the molar mass and the polymolecularity index. The mass resolving power of 12,000 with an accuracy of a few parts per million and a depth of analysis of a few tens of nanometers (matrix surface).

The mass spectrum of a molecule usually consists of a molecular peak (peak of the heaviest ion) and the strongest peak (isotope) from which the total spectrum is normalized. Other peaks correspond to different isotopes of atoms making up the molecule. They can help differentiate two molecules of the same molecular weight. Table 2.10 summarizes all the mass spectroscopy characteristics.

Incident particle	Analyzed particle	Detected chemical species	Type	Resolution	Analysis depth
UV Photon	Ionized atoms	Organic molecules	Qualitative, quantitative and destructive	Few ppm	100 nm

Table 2.10. *Summary table of mass spectroscopy*

2.4.3. Electronic analyses

Electronic analyses are defined by an interaction of the incident particle (electron or proton) with the electron cloud of the atoms constituting the target material. The interaction thus treated is that of the electron. In this section, the PIXE, SEM and EDX analyses will be presented.

2.4.3.1. X-ray analysis induced particle beams

The PIXE (particle-induced X-ray emission) analysis is a destructive surface analysis particularly suitable for the identification of elements constituting a monolayer system [JOH 95, TSU 04]. Figure 2.36 shows the principle of this method with an example of measurement of a sapphire sample.

An incident particle beam of protons of few megaelectron-volts bombards a target sample from which X-rays are emitted by ionization of the K or L layers of the atoms constituting the target (Figure 2.36(a)). An X-ray detector retrieves the X-rays retransmitted by the sample's surface. The advantage of this analysis results in a very low measurement background noise contrary to the EDX analysis using an incident electron beam. This is due to the high value of the mass of the proton with respect to that of the electron.

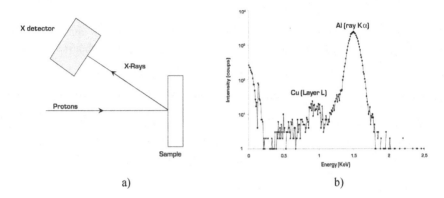

a)   b)

Figure 2.36. *a) Schematic diagram of the PIXE analysis and b) example of measurement of a sample of sapphire*

The structure studied in Figure 2.36(b) is made up of aluminum and oxygen to form sapphire (Al2O3). The Kα ray of aluminum is 1.486 keV while that of oxygen is 0.525 KeV. X-ray peaks are representative of the de-excitation of K or L electron shells of the atom. Each atom in the periodic table therefore has its specific X-ray spectrum: Kα, Kβ, etc. The Kα ray of an element is most often observed as the other lines are partially or completely masked by the noise of the detector or the spectrum of the other elements present in the sample. The effectiveness of the response of the K_α line of aluminum compared with that of the X detector is well represented on the graph in Figure 2.36b. However, the K_α line of oxygen is difficult to highlight. Oxygen is a light element and the PIXE response is lower.

Table 2.11 shows the summary of all the characteristics of the PIXE analysis.

Incident particle	Analyzed particle	Detected chemical species	Type	Resolution	Analysis depth
Protons (H^+)	X-rays (elements ionized sample)	All elements starting from boron	Quantitative and destructive	500 nm to 1 μm	0.1 to few μm

Table 2.11. *Summary table of the PIXE analysis*

The detection limits obtained are not only of the order of tens of parts per million, for most materials tested, but also highly dependent on the nature of the studied matrix.

2.4.3.2. Scanning electron microscopy

Scanning electron microscopy (SEM) is a non-destructive technique suitable for the bare chip and its assembly. It generally requires a sample preparation section with gold micro-plating (a few nanometers). It helps provide access to the dimensions of the LED structure. Figure 2.37 shows the basic principle of a SEM with the image of a GaN LED chip.

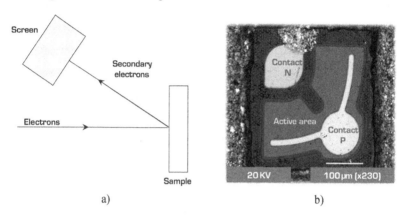

a)　　　　　　　　　　　　　　b)

Figure 2.37. *a) Schematic diagram of a SEM and b) SEM image of a GaN LED chip*

The SEM scans through an electron beam, point by point, on the surface of a sample [REI 98]. The principle of the SEM imaging is to collect the secondary electrons with an electric field of low intensity (few volts). Secondary electrons originate from a thickness less than 10 nm. The resolution of the image created on the detector is of the order of 40 Å for an incident beam diameter of 30 Å. The spot size depends on the wavelength of the electrons and hence their energy. Other particles are emitted during the electron-matter interaction: backscattered electron (Constitution analysis ranging from 0 to few micrometers), Auger electrons (surface constitution analysis) or even X-ray (electronic X-ray analysis). Table 2.12 summarizes the characteristics of a SEM.

Incident particle	Analyzed particle	Detected chemical species	Type	Resolution	Analysis depth
Electron	Secondary electrons	Materials	Quantitative, qualitative and nondestructive	500 nm	100 nm to 1 μm

Table 2.12. *Summary table of SEM*

2.4.3.3. Electronic X-ray spectroscopy

X-ray electron spectroscopy (energy-dispersive X-ray emission (EDX)) [GOL 03, NEW 86, LYM 90] qualitatively identifies the elements present on the surface of a multilayer sample. In the case of our study, this analysis is considered destructive because it requires preparation of sample by micro-section. Figure 2.38 gives a block diagram of EDX with an example of measurement of an ITO sample.

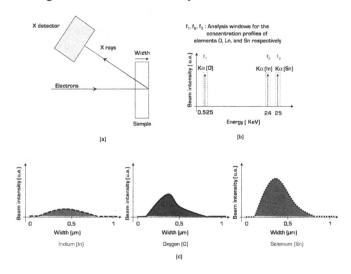

Figure 2.38. a) Schematic diagram of the EDX analysis; b) identification of analysis windows for a sample of ITO; c) profile lines of a sample of ITO

EDX analysis, associated with a SEM, consists of irradiating a sample with an electron beam. The emission that arises from the return to equilibrium is an X-ray emission. Thus, any electron from electronic heart or valence layers, whose binding energy is less than that of the incident electron, can be extracted. This identifies the type of atom and its dosage in the sample. The detection and determination can be made on items of equal or greater mass than boron (B). The spatial resolution and depth of analysis range from 100 nm to 1 μm depending on the type of material and the acceleration voltage that is chosen (Gaussian-shaped spot). Three types of measures can be extracted: X mapping of an area to be detailed, profile lines and the EDX conventional energy spectrum. In the case of Figure 2.38(c), the three elements In, Sn and O are detected at depth. The beam's intensity is based on the percentage of the element's presence. Table 2.13 provides a summary of the characteristics of EDX analysis.

Incident particle	Analyzed particle	Detected chemical species	Type	Resolution	Analysis depth
Electrons	X-rays (ionized sample elements)	All elements starting from boron	Qualitative and destructive	100 nm to 1 µm	100 nm to 1 µm

Table 2.13. *Summary table of the EDX analysis*

2.4.4. Optical analyses

Optical analyses are defined by the interaction of the incident particle (photon) with the electron cloud of atoms constituting the target material. The interaction treated is that of the photon. In this section, ATR analyses, Raman spectroscopy, fluorescence analysis and X-ray diffraction will be presented.

2.4.4.1. Attenuated total reflection

Attenuated total reflection (attenuated total reflection (ATR)) is a destructive measurement dedicated to the identification of polymer materials. It is an infrared analysis (IR) through which the optical absorption spectrum of a material is extracted in IR to determine its chemical composition. This analysis is particularly useful for coating based on organic materials, especially polymers currently used in the optoelectronics industry [VAN 06]. Figure 2.39 shows the principle of ATR [GLO 07, MIR 98, PER 05].

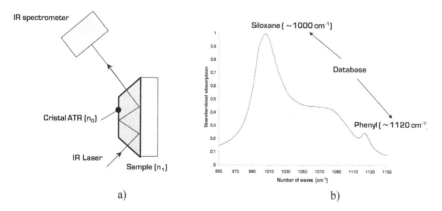

a)   b)

Figure 2.39. *a) Schematic diagram of the ATR and b) the ATR spectrum polymer coating an LED*

This analysis consists of measuring the variation in the total internal reflection of the infrared (IR) incident beam when the latter comes into contact with the sample of optical index n_1 different from that of the ATR n_0 crystal. The crystal is a highly refracting material that facilitates the IR transmission. The crystals generally used are ZnSe (500 – 1,800 nm), or Ge (800 – 1,800 nm). The IR beam is directed at an angle on a dense crystal with a high refractive index (2.38 – 4.01) [PER 05]. Internal reflection creates an evanescent wave that propagates to the sample in contact with the crystal. Some samples absorb some of the photon flux of the incident beam (IR); Evanescence will be attenuated or altered. The IR beam at the output of the crystal is recovered by an IR spectrometer detector that scans wavelength. This allows building the IR absorption spectrum of the sample as a function of wave number k (cm^{-1}). The ATR analysis is both qualitative and quantitative and has a spatial resolution of 4 – 10 cm^{-1} with a penetration depth in the sample of a few micrometers.

An example of measurement is given in Figure 2.39(b). The different molecules of the material are identified by a database of the absorption peaks of molecules as a function of the wave number. Table 2.14 gives a summary of characteristics of the ATR analysis.

Incident particle	Analyzed particle	Detected chemical species	Type	Resolution	Analysis depth
IR photon	Reflected IR photon	Polymers	Quantitative, qualitative and destructive	4 cm^{-1} to 10 cm^{-1}	Few μm

Table 2.14. *Summary table of the ATR analysis*

2.4.4.2. Raman spectroscopy

Raman spectroscopy is a non-destructive optical analysis technique (UV to IR) that is based on the Raman effect [AGI 90]. This analysis is suitable for amorphous and crystalline organic or inorganic materials (lens, electric insulation, etc.). It helps highlight the vibration modes (resonance) of molecules in order to identify the type of atoms the sample comprises [LEW 01, SMI 05, WAR 03]. The frequencies of molecular vibrations are functions of the mass of the atoms involved, and the nature of their connection. Figure 2.40 presents the block diagram of a Raman spectrometer.

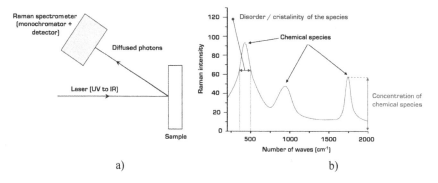

a)   b)

Figure 2.40. a) Schematic diagram of a Raman spectrometer and b) sample Raman spectrum

A laser (250 – 2,500 nm depending on the sample to be studied) is focused on the surface of the target sample. The scattered photons are focused at the entrance of the monochromator (Raman spectrometer) that constructs, through its detector, the Raman spectrum. The latter gives the light intensity (Raman intensity) of the photons scattered by the sample as a function of the wave number. The sample should be at least 100 μm thick for a spatial resolution varying from 1 to 5 μm.

The information of a Raman spectrum is as follows (Figure 2.40(b)):

– the position of the rays provides information on the chemical species present in the sample;

– the width at mid-height provides information on the disorder of the molecular structure of the sample and its crystallinity;

– the intensity of a peak can be related to the concentration of the species.

A database helps identify chemical species from their wave number. Thus, each peak of the Raman spectrum corresponds to a chemical species.

Table 2.15 shows all the characteristics of Raman spectroscopy.

Incident particle	Analyzed particle	Detected chemical species	Type	Resolution	Analysis depth
Photon (from UV to IR)	Diffused photon	Amorphous or crystalline material	Quantitative, qualitative and nondestructive	1 – 5 µm	Few 10 µm

Table 2.15. *Summary table of Raman spectroscopy*

2.4.4.3. Fluorescence spectrum

Fluorescence is a non-destructive optical analysis technique suitable for polymeric materials. It allows studying the phenomena of excitation and fluorescence emission from a sample, very common phenomena in polymers [LAK 06]. The block diagram of the fluorescence measurement is given in Figure 2.41.

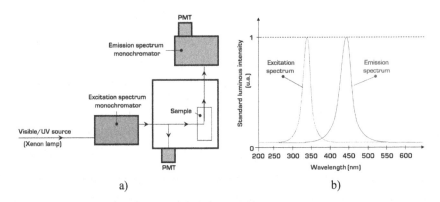

a)   b)

Figure 2.41. *a) Schematic diagram of a spectrofluorimeter and b) schematic of a measure (excitation and emission)*

An incident radiation of wavelength selected according to the sample (xenon lamp ranging from UV to visible) is focused on the sample material. A PMT (photomultiplier) detector reports the excitation wavelength spectrum. This spectrum reflects the absorption of certain molecules (fluorophores) of the sample. The peaks of the excitation spectrum thus correspond to the absorption of different fluorophores that give rise to the emission of fluorescence. The latter is focused on the input of a second monochromator in which a second PMT detector reads the sample fluorescence emission spectrum. Each peak of the emission spectrum comes from the fluorescence response of fluorophores. Thus, each fluorophore fluoresces at a particular wavelength, hence the interest in measuring in wavelength. The resolution of a spectrofluorimeter depends on the response of the sample in fluorescence. The slit width ranges from 0 to 8 mm for a spectral dispersion of 1.8 nm/mm and a light bandwidth of 0–16 nm. Table 2.16 proposes the synthesis of the characteristics of the fluorescence analysis.

Incident particle	Analyzed particle	Detected chemical species	Type	Resolution	Analysis depth
UV photon	UV/blue photon re-emitted in fluorescence	Polymer	Quantitative, nondestructive	1.8 nm/mm	-

Table 2.16. *Summary table of fluorescence analysis*

2.4.4.4. X-ray diffraction

X-ray diffraction is a structure analysis of crystallized materials. The diffraction phenomenon results from the interaction of an electromagnetic wave (X-ray) with the periodic medium of crystallized material. This analysis, nondestructive and qualitative, allows determining at the atom's or molecule's scale (few Å) the crystal orientation as well as its chemical composition. Figure 2.42 shows the block diagram of an X-ray diffractometer with a measurement example on a powder.

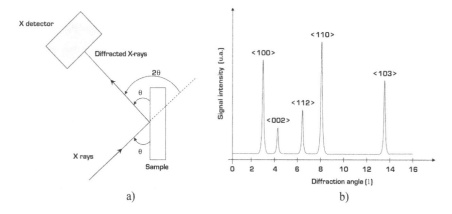

Figure 2.42. *a) Schematic diagram of an X-ray diffractometer and b) example of a diffraction pattern of a powder*

A monochromatic beam (known as $\lambda_{incident}$) of X-rays bombards the target sample with an angle respecting the law of Bragg. Indeed, the diffraction can take place in the lattice planes of the target material, if the monochromatic X-ray beam is emitted parallel and at an angle θ determined by the Bragg condition [LIF 99, WAR 90]: there must be an equality between the angle at which the X-ray beam is deflected and the angle of incidence. The emitted radiation is defined by a slit system (Soller slits) and windows located before and after the sample. The latter is spread in powder form on a glass slide which rotates with uniform motion about an axis situated in its plane (goniometric circle), thereby increasing the number of possible orientations of the lattice planes. A detector measures thereafter the intensity of the diffracted X-ray beam based on diffraction angles. It rotates about the same axis but at a speed twice that of the sample. For an angle of incidence θ, the angle measured by the meter movement will be 2 θ.

In the example in Figure 2.42(b), the identified rays correspond to crystal orientations specific to the molecules in the sample. The diffraction angle allows their determination. Table 2.17 summarizes the characteristics of the powder analysis by X-ray diffraction.

Incident particle	Analyzed particle	Detected chemical species	Type	Resolution	Analysis depth
X-ray (incident wavelength)	Diffracted X-ray (emitted wavelength)	Crystalline material	Qualitative and nondestructive	Few Å	-

Table 2.17. *Summary table of the X-ray diffraction*

2.4.5. Temperature analysis: differential scanning calorimetry

Differential scanning calorimetry: DSC is a destructive technique of thermal analysis particularly suitable for polymers and plastic materials. It allows the study of a sample's phase transitions (melting, glass transition, crystallization enthalpy ΔH, degree of reticulation, heat capacity C_p) [HÖH 03, BER 94]. The block diagram of a calorimeter is shown in Figure 2.43.

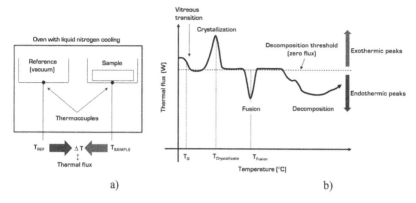

Figure 2.43. *a) Principle of DSC measurement: calorimeter.*
b) DSC measurement of a semi-crystalline polymer

The principle of the DSC is based on the measurement of heat flux required for the temperature of the sample increases or decreases. This heat flow is directly proportional to the heat capacity Cp of the material for a given temperature. Most calorimeters measure the change in this flux for temperatures between −160°C and 400°C. Some can reach 700°C with a resolution of 0.2°C.

Figure 2.43(b) shows a DSC measurement of a semi-crystalline polymer. The crystallization temperature is represented by an exothermic peak. This gives rise to heat generation; therefore, the DSC peak is above the decomposition threshold. The melting temperature, meanwhile, has an endothermic peak. In this case, there is heat absorption, and therefore, the peak is below the decomposition level.

Each phase change results in a variation in the quantity of peak heat, whose surface area is proportional to the transformation enthalpy ΔH. Melting and crystallization are thermodynamic transformations of the first order (phase change). Transformation of the second order will be characterized by a step (abrupt change of Cp) as indicated by the glass transition temperature TG of Figure 2.43(b).

Table 2.18 summarizes the characteristics of the DSC analysis.

Incident particle	Analyzed particle	Detected chemical species	Type	Resolution	Analysis depth
-	-	-	Quantitative and destructive	0.2°C	-

Table 2.18. *Summary table of the DSC analysis*

2.4.6. *Summary of physicochemical analyses*

Table 2.19 summarizes all the physicochemical analyses developed in this study for an encapsulated LED.

Type I – Bare chip			
Sample preparation	Type of analysis	Resolution	Analysis depth
Metalized bare chip on Si base	SIMS	10 – 100 Å	100 nm to few μm
Bare chip on Si base	RBS	10 – 50 nm	1 nm to 10 μm
Bare chip on Si base	PIXE	500 nm to 1 μm	0.1 to few μm
Metalized micro-section	MEB	500 nm	100 nm to 1 μm
Metalized micro-section	EDX	100 nm to 1 μm	100 nm to 1 μm
Type II – Encapsulation alone			
Sample preparation	Type of analysis	Resolution	Analysis depth
Metalized micro-section	MEB/EDX	500 nm	100 nm to 1 μm
Material alone	ATR	4 cm^{-1} to 10 cm^{-1}	Few μm
Material alone	^{1}H NMR	Few ppm	Whole sample
Material alone	Diffraction X-rays	Few Å	-
Material alone	Mass spectrometry (MALDI-TOF)	Few ppm	100 nm
Material alone (min. thickness: 100 μm)	Raman spectroscopy	1 – 5 μm	Few 10 μm
Material alone	DSC	0.2°C	-
Material alone	Fluorescence spectrum	1.8 nm/mm	-
Type III – Complete LED (chip + encapsulation)			
Sample preparation	Type of analysis	Resolution	Analysis depth
Metalized micro-section	MEB/EDX	500 nm	100 nm to 1 μm

Table 2.19. *Physicochemical characterizations by type of use (choice of analyzed area) of an LED*

Physicochemical analyses have two functions: observation of the structure of a sample and the chemical composition of different materials. The approach used in this research revolves around several axes:

– we begin by the SEM/EDX analysis to get an overview of the structure. The component is first sent to micro-section with a cutting plane allowing observation of almost all of the structure. EDX is used in order to identify the different bulk materials and pre-identify their chemical composition. However, analysis of SEM/EDX is not sufficient to identify certain non-metallic materials of the encapsulation;

– a Raman spectroscopy is used to identify inorganic materials such as the glass of a lens or the electrical insulation of an LED;

– amorphous materials may be characterized and identified by a specific ATR analysis specific to polymeric materials. ^{1}H NMR analysis and a mass spectroscopy can confirm the polymer studied. Moreover, it is not uncommon that amorphous materials are sensitive to certain phenomena associated with the interaction with the LED's light and the temperature. The fluorescence analysis and Raman spectroscopy are used for the study of these phenomena;

– regarding the crystallized material (YAG: Ce phosphor powder), X-ray diffraction analysis is particularly well adapted;

– according to areas of interest, more detailed analyses can be performed. Indeed, to build electrical and optical models related to the technology of the bare chip, it is necessary to know the composition of each epitaxial layer and its doping level. Comparing the results of SIMS, RBS/PIXE and SEM/EDX analyses provides such information;

– we end with the degradation analysis of the component materials. ^{1}H NMR analysis, mass spectrometry, ATR, XRD, DSC and fluorescence spectrum are adapted to the analysis of the degradation of the assembly, while the SIMS, RBS/PIXE and SEM/EDX analyses are more appropriate for chip dimensions.

The usage approach will be shown in Chapters 3 and 4.

2.5. Conclusion

In this chapter, four main phases in the methodology of failure analysis of this research have been described.

The first phase gives the architecture of the methodology used for the measurement of the junction temperature. Since the functional and physical parameters are T_j dependent and that only T_P during the manipulation, it is important to develop methods to determine T_J. Two methods have been proposed in this section to determine the junction temperature of an optoelectronic component:

– an optical method, conducted through a monochromator, which implies that we know the component of the active region structure and its physical properties;

– an electrical method, using a laser power coupled to a pulse generator, requiring no knowledge of the physical properties of the component but may be limited by the accuracy of measurement depending on the apparatus used (e.g. oscilloscope). The measurement can be accurate to ±2–5 K depending on the type of oscilloscope.

These two methods are complementary and produce the same result: the characterization of the component studied, T_J, is known to build physical models at a temperature of the active area of the known component.

Sections 2.2 and 2.3 describe those of a MQW InGaN/GaN structure. These models are extracted from the electrical characteristics I(V) realized from a femtoampere meter. The measurement of the optical power was carried out using a UV photodiode connected to an OPHIR displaying the values. The spectral characteristics were identified through the monochromator. All of the electrical and optical measurements were carried out at constant room temperature regulated by a liquid nitrogen cryostat controlled by a temperature regulator. It is emphasized that the GaN material imposes a P contact of Schottky type. The models from this chapter will serve as tools that we will use to establish the failure analysis of low- and high-power LEDs in Chapters 3 and 4, respectively.

The physicochemical analyses defined in section 2.4 offer a broad spectrum for the identification of materials constituting the complete LED and their chemical composition. They also help to help build optical models of LEDs and study degradation mechanisms, when a component is subject to environmental constraints, explaining changes in physicochemical parameters of each material studied.

The methodology of failure analysis will be presented in Chapter 3, based on physical models presented in this chapter. This methodology for GaN (blue) LEDs has an encapsulation suitable for space applications. The tools introduced in this chapter to study the impact of aging in active storage (1,500 h/85°C/I rated) on each technology will be used.

3

Failure Analysis
Methodology of Blue Leds

Chapter 2 focused on the description of the basic tools required for failure analysis of optoelectronic components for an InGaN/GaN MQW structure. In particular, we have presented two methods (electrical and optical) for measuring the junction temperature. We have shown that this parameter, as well as information provided by the physicochemical analyses, represents major points in the preparation of electrical and optical models of both types of the studied structures. Thermal, electro-optical and physicochemical analyses have demonstrated the wealth of information provided in order to develop a complete physical model.

Increasing the integration of optical and electronic functions in one same assembly necessitates the establishment of appropriate methodologies, based on simple but complete models for the evaluation of all the physical mechanisms involved in the degradation of a component subjected to its environment.

Chapter 3 will help to highlight the range of possibilities electrical, optical and thermal analysis benches developed at IMS Laboratory, both in terms of performance for the detailed characterization of the technologies, and at the level of interpretations for failure analysis. This work will be divided into five separate points:

– the definition of environmental stresses;

– the aging campaign carried out as part of the project. This section focuses on aging carried out in this research (active storage) and the steps needed to prepare the failure analysis;

– the objective of the third section is to present the electrical and optical models, built from the physical mechanisms detailed in Chapter 2, precisely and specifically adapted to the studied structure. The components as well as their assembly will be presented;

– the implementation and progress of the methodology of failure analysis. The objective of this section is to highlight the electrical and optical failure signatures that will enable pre-locating sensitive areas, both in terms of the component and its assembly. The physicochemical analyses will, if necessary, confirm the location of the degraded areas and the failure mechanism(s) inducing the component's degradation;

– a summary of the activities developed in this chapter will be presented in the conclusion.

The purpose of this chapter is to show that the methodology is adaptable to the failure analysis of already marketed low-power components (<30 mW) under aging determined from the components' functional operating conditions.

3.1. Mission and aging profile

3.1.1. Component definition

These stresses therefore require:

– a sealed box with space radiation (gamma rays and protons);

– a JANTXV (Joint Army–Navy Technical Exchange Visual Inspection) certification ensuring a level of reliability satisfying the requirements of the space field;

– an operating temperature ranging from –20°C to +85°C.

In this context, the selected components were manufactured by OPTRANS and their details are given in Table 3.1.

Reference	Active area	P_{opt} (mW)	T_{min}/T_{max} (°C)	λ_C (nm)	$I_{nominal}$ (mA)	$\Theta_{emission}$ (°)	Number of LEDs
VS472N	MQW InGaN/GaN	1.4 at 20 mA	– 20/+85	472	30	12	75

Table 3.1. *Selected component data derived from OPTRANS*

P_{opt}, optical power; T_{min}, minimum operating temperature; T_{max}, maximum operating temperature; λ_C, central wavelength; I_{rated}, rated (and maximal) current and $\theta_{emission}$, emission angle.

3.1.2. Environmental stresses and acceleration factor

For optoelectronic components mainly composed of plastic and metal materials such as LEDs, environmental stresses likely to damage the device are mainly temperature and supply current [FUK 91, UED 96]. Plastic materials, increasingly used in the optoelectronic industry, limit the component's operating temperature. Indeed, it is quite rare to exceed an external temperature of 100°C with this type of component. The disparity of assembly materials usually leads to the application of thermal aging tests [DES 02]. This type of test allows us to understand the robustness of an assembly and is therefore subject to MIL (military)-type standards. In our case, we are interested in the component's behavior during a mission: it is an operational aging test. Accordingly, fixed-temperature aging with or without power is carried out in a laboratory. This type of aging corresponds to active or passive storage. Generally, it is based on the temperature and the maximum operating current specified by the component's technical documentation. We can, in certain cases, apply an acceleration factor in current and/or temperature. The acceleration factor temperature A_{FT} generally follows an Arrhenius law [DES 02, UED 96, WAD 94]. It is expressed by equation [3.1]:

$$A_{FT} = \exp\left[\frac{E_a}{k}\left(\frac{1}{T_{op}} - \frac{1}{T_{acc}}\right)\right]$$
[3.1]

E_a is the activation energy, T_{op} is the operational temperature and T_{acc} is the temperature in accelerated conditions.

The acceleration factor, based on experimental measurements [DES 04], in current A_I generally follows a power law described by equation [3.2]:

$$A_I = \left(\frac{I_{acc}}{I_{op}}\right)^{\beta}$$
[3.2]

β is the power parameter, I_{acc} is the current in accelerated conditions and I_{op} is the operational current.

Increasing the power density within the existing optoelectronic components induces self-heating of the component when the latter is powered. The consequence of the increase in current is therefore an increase in temperature. The final acceleration factor A_F is written, therefore, in the form of equation [3.3].

$$A_F = \left(\frac{I}{I_{op}}\right)^{\beta} \exp\left[\frac{E_a}{k}\left(\frac{1}{T_{op}} - \frac{1}{T_{ae}}\right)\right] = A_{FT}.A_I \qquad [3.3]$$

T_{ae} is the temperature of the component with self-heating.

This first analysis allows justifying the manipulation in order to determine a chip's junction temperature. Even under operational conditions, the chip's temperature is higher than that of the outer packaging. We thus have a thermal model that provides access to the estimation of the junction temperature for a packaging temperature T_p and a given current I (see Chapter 2).

3.2. Aging campaigns

Aging campaigns are defined by specifications corresponding to the needs related to the components' field of application. This section focuses on aging conducted in this study (active storage) and the steps necessary for preparing failure analysis.

3.2.1. Specifications of accelerated aging

The specifications were elaborated by the IMS laboratory. Active storage is storage in fixed temperature and current supply. Table 3.2 summarizes the entire campaign initiated on three LEDs.

LED technology	Aging temperature (°C)	Supply current (mA)	Aging time (h)	Number of aged LEDs
GaN – VS472N	85	30	1,500	3

Table 3.2. *Aging campaign in active storage*

The aging campaign was conducted in the IMS laboratory using HERAEUS Instrument-type ovens with current laser supplies of PROFILE PRO8000 type. The

LEDs are connected to a system of DIL (dual in line) type and are not connected to a heat sink. Many electrical and optical characterizations were measured:

– at 0, 1,000 and 1,500 h: the characteristics I (V), P (I) and L (E) were measured for the packaging temperatures of 243, 273, 300 and 373 K. The far field P (θ) was taken at a packaging temperature of 300 K. Further measurements of P (I) and L (E) were carried out for a 300 K junction temperature;

– at 168 and 500 h: the characteristics I (V), P (I), L (E) and P (θ) were carried out for a packaging temperature of 300 K only. Additional measurements P (I) and L (E) were carried out for a 300 K junction temperature.

A sample LED per technology was characterized at room temperature before the start of each electrical and optical characterization in order to verify the stability of the measurement bench.

3.2.2. Aging campaign

This section presents the progress of an aging campaign. The three major points discussed are:

– the metrological aspects ensuring the measurements' repeatability and reproducibility;

– the electro-optical analyses necessary for the verification of the electro-optical parameters given by the manufacturer;

– the physicochemical analyses allowing the adaption of the electrical and optical models identified from the information on the studied component's materials.

3.2.2.1. Electrical and optical measurement bench: metrological aspects

During an aging campaign, it is necessary to establish the metrology of the measurement benches that are used. This ensures the repeatability and reproducibility of measurements and quantifies the uncertainties of each parameter extracted from electrical and optical measurements. This step thus gives us the assurance that the observed variations in the electrical and optical parameters are due to disturbances caused by the aging test. It takes place in two phases:

– a preliminary phase of the bench's metrology before starting the initial measurements;

– a phase present at each stage of aging (retaking measurements) which checks, from a sample component (new), the bench's stability and measurement uncertainties;

The metrology of a measurement bench is defined with three key terms[1]: precision, correctness and accuracy of a measurement method.

Precision refers to the fineness of the agreement between the different test results. The need to consider the precision arises because tests performed on identical components in identical alleged circumstances do not give, in general, identical results. This is attributed to unavoidable random errors inherent in any measurement procedure. Factors that influence the result of a measurement cannot all be fully controlled. In the practical interpretation of measurement data, this variability should be taken into account. For example, the difference between a test result and a specified value can be found within the inevitable random errors, in which case, areal deviation from the specified value has not been established. Similarly, the comparison of test results from two lots of materials will not indicate a fundamental difference in quality if the difference between them can be attributed to an inherent variation in the measurement procedure. Precision therefore has two extremes: repeatability and reproducibility. The repeatability gives the minimum of variability in the results and reproducibility gives the maximum variability in the results.

The trueness of a measurement method is of interest when it is possible to conceive a real value for the measured property. Although, for some measurement methods, the real value is not known exactly, it is possible to have an accepted reference value for the measured property. The trueness of a measurement method can then be investigated by comparing the accepted reference value with the level of results given by the measurement method.

Accuracy is the combination of precision and trueness. It is convenient to determine the measurements of accuracy from a series of test results, reported by the participating laboratories organized under the guidance of an expert panel that is specifically established for this purpose.

Thus, for each measurement bench associated with aging, measurement devices have been verified by a component used as a reference by characterizing the same component for the same operator. For example, for the characteristics I(V) of aging in active storage, an LED per technology was characterized (same operator) three times with assembly and disassembly of the LEDs before each measurement. An example of an electrical measurement on a GaAs LED is shown in Figure 3.1.

---

1 NF ISO 5725-1: definitions of precision, trueness and accuracy according to the ISO/DIS 3534-2 standard.

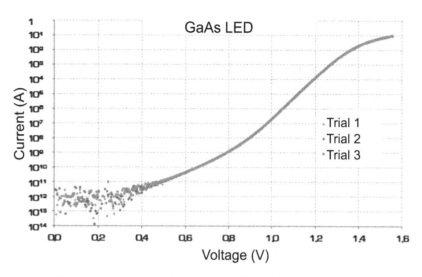

Figure 3.1. *Metrology test: characteristics I(V) of a GaAs LED assembled and disassembled three times (Keithley 6430)*

Sensitive parameters tested for the validation of electro-optical measurement benches presented in Chapter 2 are listed in Table 3.3.

Measurement bench	Characterization type	Packaging temperature (T_p)	Parameter	Values obtained for three trials	Error (%)
I(V)	Electrical	300 K	R_s (Ω)	1.776/1.775/1.775	0.056
			V_{th} (V)	1.378/1.378/1.379	0.073
T_J(I)	Electrical	300 K	T_J (K)	315.18/317.64/317.77	0.815
P(I)	Optical	300 K	P_{opt} (mW)	24.81/24.73/24.82	0.363
L(E)	Optical	300 K	λ_C (nm)	875.24/875.32/875.17	0.017
			$\Delta\lambda$ (nm)	41.54/41.47/41.57	0.241

Table 3.3. *Sensitive electrical and optical parameters of GaAs LED powered at 100 mA allowing the electrical and optical measurements bench's validation*

For the sake of clarity and simplification, the error bar was not added to the measurements presented in this study. Drifts presented in this book reflect measurement errors caused by the devices.

3.2.2.2. Electro-optical analyses: verification of electrical and optical parameters given by the manufacturer

An important point in a reliability analysis is to compare the measurements made by the laboratory to those indicated by the manufacturer. Table 3.4 compares the typical values of the main electrical and optical parameters given by the manufacturer OPTRANS with those measured in the IMS laboratory.

LED technology and conditions	Parameter	Symbol	Typical value OPTRANS	Measured value IMS
GaN – VS472N I = 20 mA T_p = 300 K	Optical power (mW)	P_{opt}	1.4	2.2
	Direct voltage (V)	V_F	3.5	3.22
	Central wavelength (nm)	λ_C	468	467
	Width at mid-height (nm)	$\Delta\lambda$	35	24
	Emission angle (°)	θ	12	12

Table 3.4. *OPTRANS/IMS typical value comparison*

The differences noted between the two columns OPTRANS/IMS are due to the number of measured components. A manufacturer guarantees a typical value with a certain safety margin compared with the actual value of the parameter ($\approx$20%). To satisfy the client, the latter will always give a pessimistic value compared with actual measurements. This ensures the client the device operation in the chosen application because it is based on typical values to build the component. On the other hand, the manufacturer's measuring instruments are designed to characterize thousands of components (production lines). The IMS laboratory test bench is, in turn, adapted to measurements of some components by campaign.

Finally, the verification of measured electrical and optical parameters allows building and validating electro-optical models used from the materials' information (size, type, doping levels, etc.). This information can be given by the manufacturer or extracted from different physicochemical analyses.

3.2.2.3. Analyses of the structure and materials of the encapsulated component

Physicochemical analyses are tools used to illustrate the structure and materials constituting the device. This step validates the electro-optical models by identifying the different types of materials present in the encapsulated LED, the dimensions of the different parts of the component (chip, bonding, lens, packaging, etc.) and sometimes the emitting chip's layers' doping levels.

We thus use the physics of the component giving electrical and optical theoretical models and compare them with experimental measurements. In this manner, the extracted model of the electro-optical analyses connects the areas of the studied structure to the electro-optical characteristics. Table 3.5 summarizes all the physicochemical analyses conducted on the two types of LEDs.

LED technology	Analyzed part	Analysis conducted	Number of analyzed LEDs
GaN – VS472N	Complete chip	MEB/EDX	2
	Complete chip	MEB/EDX	2
	Complete chip	MEB/EDX	2
	Complete chip	SIMS	8
	Chip :active area and contacts	RBS/PIXE	2
Assembly	Polymer coating	ATR	2
		RMN ^{1}H	
		Mass spectrometry	
	Polymer coating	DSC	5
	Electrical isolator	Raman spectroscopy	1
	Complete package	MEB/EDX	2

Table 3.5. *Summary of physicochemical analyses conducted in this study*

3.3. Initial characterization of LEDs: electrical and optical aspects

Initial characterization of LEDs used to establish electrical and optical models constitutes a necessary basis for failure analysis. The objective of this section is to present the electrical and optical models, built from the detailed physical mechanisms in Chapter 2, adapted to the studied structures: GaN MQWLED. We will see that these components' assembly can play a major role in optical modeling.

3.3.1. Leds' technological description

3.3.1.1. $In_xGa_{1-x}N/GaN$ multi-quantum well (MQW) LEDs

Figure 3.2 illustrates a cross-sectional view (micro-section) of the studied GaN LEDs' chip.

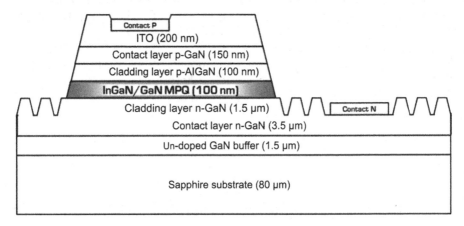

Figure 3.2. Cross-sectional *diagram of an InGaN/GaN MQW LED's chip with its various layers*

This structure consists of a thick sapphire substrate, with a thickness of 80 µm, which is deposited on an un-doped transition layer of GaN, with a thickness of 1.5 µm. Above the latter, an N-doped GaN layer $(Si/10^{17}–10^{19}$ cm$^{-3})$ with 5 µm thickness etched by wet etching or plasma ICP (inductively coupled plasma) etching, with studs of 5 µmin diameter and 200 nm thickness, is deposited. This layer serves as a contact layer on which is deposited the N contact composed of a semi-transparent conductive oxide – rhodium oxide $(Rh_2O$ – thickness of 200 nm). The active region comprises three InGaN/GaN quantum wells (100 nm in total) above which an AlGaN P $(Mg/10^{18}$ cm$^{-3})$-doped confinement layer (100 nm) is found. The P contact (Rh_2O), with a thickness of 200 nm, is deposited on an indium oxide layer doped with tin (ITO: indium tin oxide), itself deposited on a P $(Mg/4.45.10^{18}$ cm$^{-3})$-doped GaN (150 nm) layer separating the ITO layer from the P AlGaN layer.

SIMS analyses (INSA Toulouse) correlated with RBS/PIXE (Laboratory CENBG) and SEM/EDX (IMS Laboratory) analyses have determined the structure shown in Figure 3.2.

3.3.1.2. Structure of the packaging

The packaging of a "bare chip" is the terminal phase in the development of a commercial component. The various assemblies allow to:

– guide the light of the bare chip to the application or photonic system;

– dissipate heat from the chip during operation;

– protect the chip from environmental constraints.

The packaging, of T047 type, is the same for both types of LEDs. It is adapted to the environmental constraints imposed by the spatial field. Figure 3.3 presents a cross-sectional view of the structure of the encapsulating packaging.

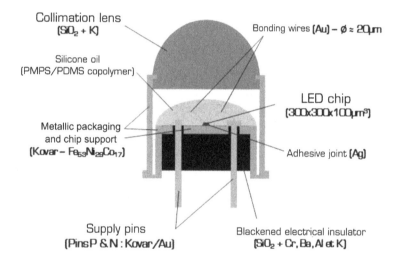

Figure 3.3. *Cross-sectional diagram of the encapsulating packaging of OPTRANS LEDs*

The base of the packaging is made of Kovar ($Fe_{53}Ni_{29}Co_{17}$), wherein a transparent silicone oil (copolymer: $PMPS^2/PDMS^3$) coats the emission chip. ATR and NMR ^1H analyses allowed determining the chemical composition of the copolymer. Figure 3.4 shows the comparison of the ATR spectra of un-aged samples of PMPS, PDMS and studied silicone oil.

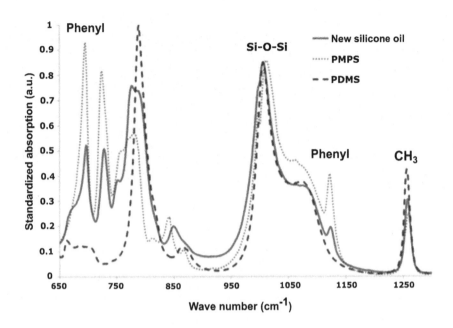

Figure 3.4. *Comparison of ATR spectra of three new samples: PMPS, PDMS and studied silicone oil*

This coating acts as protection of bonding wires and chip against chemical attack and external vibrations.

[2]Polymethylphenylsiloxane

[3]Polydimethylsiloxane

It also improves the output of LED and has a transmittance greater than 80% in the visible range (see Chapter 1).

On the same optical axis, a collimator glass (SiO_2) lens is arranged to reduce the emission angle to 12°. SEM/EDX analyses complemented by an IR-type Raman spectroscopy analysis identify the composition of glass. The glass is therefore doped with potassium, thereby reducing the melting point temperature from 1,800 to 1,400°C. This doping and thermal process increase the hardness of glass (>520: Class 6 on Knoop's scale) [GAN 06, OPT].

Light is emitted by the chip ($300 \times 300 \times 100$ μm^3) attached to the metallic (Kovar/Au) support by an adhesive joint loaded with silver (Ag: 75%) [ADA 02].

The latter includes (Kovar/Au) power pins coming into contact on the chip. The two contacts (N and P) are taken on the upper side of the emitting chip.

A SEM/EDX analysis was used to verify the composition of the packaging, metallic mount, adhesive seal, bonding wires and the power pins.

The electrical insulation of the power connectors is performed by a highly doped glass allowing light absorption: "blackened glass" (SiO_2 + Cr, Ba, Al and K). This material is used to prevent the emission of light by the rear face of the packaging. This glass was identified by a Raman spectroscopy coupled to a SEM/EDX analysis.

All collaborations engaged with national laboratories for physicochemical analyses having confirmed the information materials described above on the chips and their assembly are summarized in Table 3.6.

Laboratory	Conducted physicochemical analyses	Concerned area of LED
IMS – Bordeaux	SEM/EDX	Whole LED (chip and packaging)
ICMCB – Bordeaux	Raman spectroscopy	Lens and electrical insulator
CENBG – Bordeaux	RBS/PIXE	Chip
INSA – Toulouse	SIMS	Chip
ISM – Bordeaux	ATR RMN ^{1}H	Silicone oil

Table 3.6. *Physicochemical analyses and collaborations*

3.3.2. Extraction of LEDs' electro-optical parameters

This section is central to the methodology for analyzing the physics of failures. Indeed, it is based on Chapter 2, describing the theory of transport phenomena and electronic transitions in an LED. This section shows the establishment of an electro-optical model based not only on electro-optical measurements, but also on physicochemical analyses. In this way, each electrical and optical signature is clearly associated with each part of the component's structure.

The failure signature analysis will be simplified and will help guide the physicochemical analyses to validate the established failure model.

3.3.2.1. *LED's equivalent electrical models and typical values*

Figure 3.5 shows the characteristic I(V) typical of MQW InGaN/GaN LED with the various working zones associated with their respective electrical parameters.

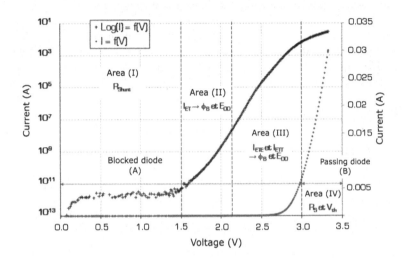

Figure 3.5. *Typical electric characteristic of an InGaN/GaN MWQ LED*

Figure 3.6 illustrates the equivalent electrical model of a MQW InGaN/GaN LED.

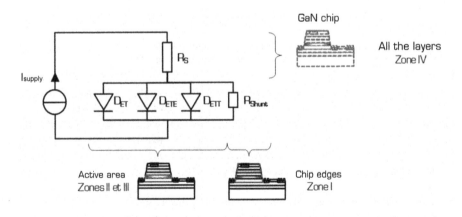

Figure 3.6. *a) Equivalent circuit diagram of an InGaN/GaN MQW LED and b) location of the parameters on the GaN chip based on the operating zones (I–IV)*

The D_{FE}, D_{TFE} and D_{TE} diodes correspond to currents FE, TFE and TE, respectively. The typical values of the electrical parameters are provided in Table 3.7.

Parameter	Equivalent model	Concerned zone	Extraction curve	Typical value of an InGaN/GaN MQW LED at $T_J = 300$ K and I = 30 mA	References
R_{Shunt} (Ω)	R_{Shunt}	I	$Log(I) = f(V)$	$4.10^{11} - 7.\,10^{11}$ Ω	[MOR 08c, MOR 08b, DES 10]
φ_B(eV)	D_{ET}, D_{ETE} and D_{ETT}	II and III	$Log(I) = f(V)$	$1.6 - 1.7$ eV	[MOR 08c, MOR 08b, DES 10]
E_{00}(meV)	D_{ET}, D_{ETE} and D_{ETT}	II and III	$Log(I) = f(V)$	$80 - 100$ meV	[MOR 08c, MOR 08b, DES 10]
V_{th}(V)	V_{th}	IV	$I = f(V)$	$1.82 - 1.83$ V	[MOR 08c, MOR 08b, DES 10]
R_S (Ω)	R_S	IV	$I = f(V)$	$8.3 - 8.6$ Ω	[MOR 08c, MOR 08b, DES 10]

Table 3.7. *Typical values of the electrical parameters of an InGaN/GaN LED*

3.3.2.2. LED's optical models and typical values

Theoretical optical models of a MQW GaN LED structure are based on three mechanisms of electronic transitions (spontaneous emission, stark effect and optical gain).

Table 3.8 outlines the typical values of the optical parameters related to the MQW structure of a GaN LED's chip.

Parameter	Extraction curve	Typical value of an InGaN/GaN MQW LED: $T_J = 300$ K and I = 30 mA	References
K_{Spon} $(cm^{-1}.eV^{-1/2})$	Rspon (hv)	$10^{-4} - 10^{-3}$	[ROS 02, DES 02, DES 10]
τ_R (s)	Log(I) = f(V)	10^{-9}	
γ_{max} (cm^{-1})	γ (hv)	$10^5 - 10^6$	
χ_{VC} (m)	-	$0.4 - 0.5$ Å	
m	Rspon (hv)	$1.1 - 1.2$	
ΔE_F (eV)	Rspon (hv)	$2.9 - 3.1$	
α_{2D} (m^{-1})	γ (hv)	$10^{-5} - 5.10^{-5}$	
ε_{C1} (meV)	γ (hv)	150	
ε_{V1} (meV)	γ (hv)	-40	
β (meV)	R_S (ξ)	$60 - 180$	
r_0	R_S (ξ)	$450 - 550$	
F $(V.m^{-1})$	R_S (ξ)	$10^7 - 10^8$	

Table 3.8. *Typical values of the optical parameters of the MQW structure of an LED InGaN/GaN chip*

Figure 3.7 compares the theoretical models to the experimental measurement of the standard optical spectrum of a MQW GaN LED.

We notice a deformation of the experimental optical spectrum when compared with the theoretical spectrum. The theoretical model takes into account the fact that GaN's conduction and valence bands are non-parabolic, and the gain in the quantum well is almost constant. The peaks observed at lower and greater energies than that of the LED light are not caused by this phenomenon. The theoretical spectrum only takes into account the physical mechanisms involved in the emitting chip. However, the experimental spectrum is a measurement of the light emitted from the assembled

LED. Therefore, it is a phenomenon outside the emitting chip that distorts and enlarges the experimental spectrum.

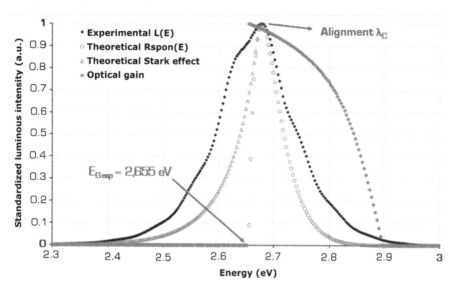

Figure 3.7. *Standard models of a GaN MQW LED optical spectrum*

For wavelengths approaching the blue and UV (case of GaN Leds), a phenomenon, very common in the polymeric materials, may intervene: the silicone oil fluorescence emission [DIA 01, MAC 91, KRE 97]. The hypothesis that we can thus formulate is that the chip's light is partially absorbed by the silicone oil which re-emits in fluorescence.

To confirm this fluorescence hypothesis, the silicone oil present in the assembly of the LEDs studied in this chapter was excited at two specific wavelengths:

– at 360 nm: this allows verifying the oil's sensitivity to UV radiation;

– at 464 nm: this wavelength is the central wavelength of the LED chip. This allows verifying the emission of fluorescence when the silicone oil is excited by the light from the LED (actual conditions).

Figure 3.8 shows the fluorescence emission spectrum of a new sample of oil silicone excited at 360 nm and 464 nm.

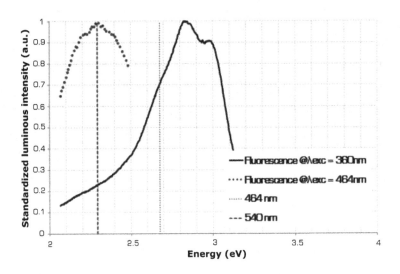

Figure 3.8. *Fluorescence emission spectra of a new sample of oil silicone excited at 360 nm and 464 nm*

The fluorescence spectrum of the silicone oil, excited in the UV (360 nm), confirms the emission of fluorescence to the corresponding energies to those of the light emitted by the LED (2.672 eV→464 nm). When the coating polymer is excited at the wavelength of the light emitted by the chip (464 nm) fluorescence, emission appears with an emission peak at 540 nm. This validates the assumption made by the analysis of optical spectra and is used to locate the physical phenomenon explaining the deformation of the GaN spectrum.

To explain the physical origin of the fluorescence emission and build the corresponding photonic models, several hypotheses have been proposed according to energy areas:

– For energy area greater than that of the LED's emission (2.67 eV), three hypotheses were formulated:

- hypothesis 1: the light emitted by the chip is absorbed by the silicone oil in the form of two-photon absorption, which would require nonlinearity in the silicone oil's absorption spectrum;

- hypothesis 2: The anti-Stokes Raman phenomenon is effective in silicone oil. The vibrations of the molecules will then provide a sufficient energy gain to transmit at an energy higher than 2.67 eV;

- hypothesis 3: a phenomenon of absorption/re-emission of fluorescence is effective and deforms the spectrum at energies higher and lower than that of the LED's emission (2.67 eV).

Photonic models of the three hypotheses are presented in Figure 3.9. The silicone oil is modeled by an excited singlet state S1 and a fundamental singlet state S_0, both at several vibration levels. For the two-photon absorption and Raman anti-Stokes models, a theoretical virtual state is introduced in order to clarify the phenomenon.

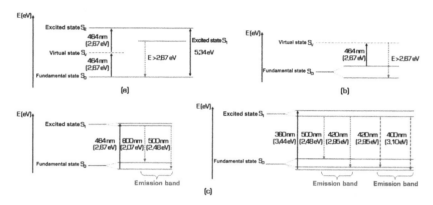

Figure 3.9. *Jablonski diagrams of the silicone oil for the three hypotheses: a) two-photon absorption; b) Raman anti-Stokes phenomenon; c) fluorescence absorption/re-emission process*

The two-photon absorption and the Raman anti-Stokes mechanism are located on the experimental optical spectrum of the LED in Figure 3.10.

It is possible for an atom or a molecule to "simultaneously" absorb two photons (Figure 3.9(a)). The two-photon absorption is thus an optical transition between two quantum states involving an almost simultaneous presence of two photons. The atom is first de-energized non-radiatively to an intermediate energy state S_1 and then de-energized to its fundamental level S_0 by emitting a photon of energy greater than the energy of the incident photons. This phenomenon is the fluorescence by de-excitation of states created by two-photon excitation. The actual performances of avalanche detectors based on a semiconductor (Si or GaAs) recently allow detecting these unlikely events. Indeed, the used light intensity must be very large (>> Power density 10^3 W.cm^{-2}), so that the probability of two-photon absorption is sufficiently large to be observable. Pulses of very short duration (<200 fs) are needed to achieve sufficient intensity for the observation of this phenomenon [DUF 06]. This effect is due to nonlinear properties that exist at the focal point of the target material (silicone oil) due to the very high energy concentration. The hypothesis of the two-photon absorption is therefore only valid if nonlinearities in the absorption spectrum of the silicone oil are assumed. The power density emitted by the LEDs, being too low (2.9 W.cm^{-2}), can rule this hypothesis.

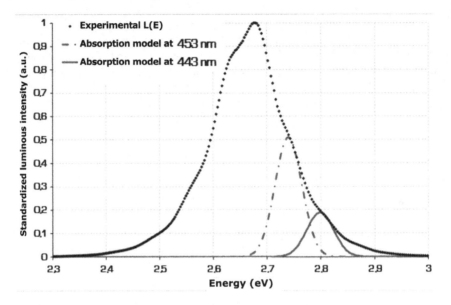

Figure 3.10. *A GaN LED's optical spectrum: localization of the phenomena of two photon absorption and Raman anti-Stokes*

The second hypothesis refers to the anti-Stokes Raman effect located at the same place as the two-photon absorption hypothesis (Figure 3.10). In quantum mechanics, Raman scattering is described as the excitation of a virtual state, lower in energy than a real electronic transition and its "de-excitation" to a real vibrational state of the fundamental state. Such a diffusion event occurs in less than 10^{-14} s [GAL 06]. At room temperature, a small fraction of the molecules of the studied material are in vibrationally excited states, and the Raman scattering from these vibrationally excited molecules takes them back to their fundamental state. Under these conditions, the scattered photon appears with a higher energy than the excitation photon as shown in Figure 3.9(b). Thus, the incident photon draws energy from molecules to bring them to a lower vibrational level, in this case from S_V to the lowest level S_0. The Raman anti-Stokes effect is visible in the ranges of UV up to IR. Now, in the UV, fluorescence of polymeric materials may be much more intense than the Raman Effect to the point of completely hiding it. The use of NIR lasers can limit, if not, eliminating the harmful effects of fluorescence. Our point of interest is in the range of blue/UV, which helped rule out the anti-Stokes Raman phenomenon, which is generally very low in light intensity compared with the fluorescence and does not explain a 20% (443 nm) to 50% (453 nm) fluorescence emission in the LED's experimental optical spectrum (Figure 3.10).

The fluorescence absorption/re-emission phenomenon is therefore the most probable hypothesis for such important fluorescence intensities such as those present in the LED's optical spectrum. Figure 3.11 locates fluorescence absorption/re-emission areas for energies superior and inferior to those of GaN LED's central wavelength.

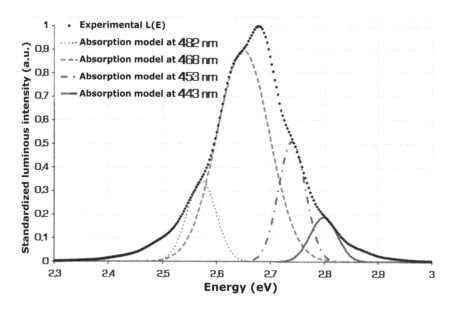

Figure 3.11. *Optical spectrum of a GaN LED: localization of absorption/re-emission phenomena*

To confirm this hypothesis, the silicone oil's fluorescence study was continued by measuring the absorption spectrum at an observation wavelength corresponding to the maximum (540 nm) of the fluorescent emission of the oil excited at 464 nm. Figure 3.12 shows the fluorescence emission spectra of the new silicone oil sample excited at 360 nm and 464 nm, by adding the absorption spectrum at 540 nm.

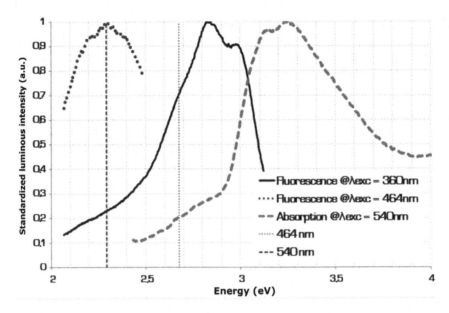

Figure 3.12. *Fluorescence emission spectra of a new silicone oil sample, excited at 360 nm and 464 nm, superposed on the absorption spectrum of the excited fluorophores of the oil excited at 540 nm*

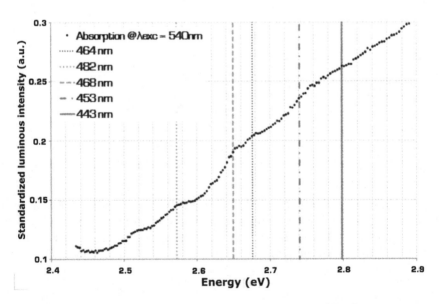

Figure 3.13. *Zoom on the energy window (about 464 nm) of the fluorescence absorption spectrum of a new silicone oil sample excited at 540 nm*

The symmetry between the emission spectrum at 360 nm and the absorption spectrum of the fluorophores at 540 nm confirms the process of fluorescence absorption/re-emission. Moreover, by focusing on the energy area corresponding to the absorption process at energies below and above that of the LED's central wavelength (464 nm, 2.67 eV), we find the four absorption peaks. Figure 3.13 shows the energy window of the absorption spectrum at 540 nm, indicating the absorbed wavelengths.

This confirms the optical absorption models of Figure 3.11 as well as the hypothesis of fluorescence absorption/re-emission by the fluorophores. The photonic model is proposed in Figure 3.9(c) with two fluorescence emissions:

– 464 nm excitation: emission of fluorescence of 500–600 nm with a maximum at 540 nm.

– UV excitation (360 nm): fluorescence emission from 400 to 420 nm with a peak at 415 and 420–500 nm with peaks at 435, 443, 453, 464, 468 and 482 nm.

Thanks to the excitation spectrum and emission spectra, we can calculate the fluorescence yield defined as the ratio between the number of emitted photons and the number of photons absorbed by the fluorophore molecules [LAK 06]. This ratio is generally between 0.05 and 1. Its expression is given by equation [3.4].

$$\eta_{\text{fluorescence}} = \frac{I_{\text{fluo}}}{I_{\text{abs}}} \tag{3.4}$$

Table 3.9 shows the values of the fluorescence yield of the studied silicone oil designed for fluorescence at both analyzed excitations.

Excitation wavelength (nm)	Fluorescence peak (nm)	$\eta_{\text{fluorescence}}$ (%)
360	415	27.32
	435	29.74
	443	28.94
	453	24.51
	464	21.01
	468	19.49
	482	14.63
464	540	99.16

Table 3.9. *Values of the fluorescence yield of the silicone oil excited at 360 nm and 464 nm*

An excellent fluorescence yield (> 99%) is observed when the silicone oil is excited by light from the LED, which is at 464 nm.

To explain the origin of the fluorescence absorption/re-emission mechanism, an RMN ^{1}H analysis and mass spectrometry were performed. The purpose of these physicochemical analyses is to detect the presence of fluorophore molecules responsible for absorption of the LED light and its re-emission by fluorescence. Figure 3.14 shows the RMN ^{1}H spectrum and the mass spectrum of a new silicone oil sample.

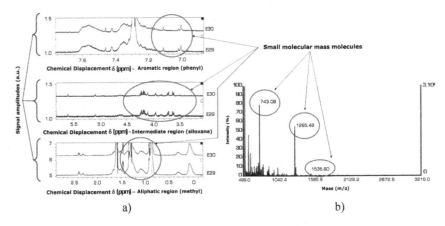

a)   b)

Figure 3.14. a) RMN spectra of the various regions of two silicone oil samples (S29 and S30) extracted from the assembly of GaN LEDs and b) mass spectrum of sample S29

The RMN spectra have allowed locating the presence of low-molecular-weight molecules (Small Molecules: SM) in the various regions corresponding to phenyl, methyl and siloxane patterns of the studied silicone oil. The circled areas define the peaks corresponding to the SM. Their presence was confirmed by a mass spectrum with three peaks indicating the SM. The origin of the fluorescence absorption/re-emission mechanism is the existence of SM acting as fluorophores.

This phenomenon of fluorescence shows the assembly's impact on the LED's optical properties: a notable difference is remarkable between the single chip's emission and the emission measured at the LED assembly's output. In the next section, we will see the evolution of this impact when the component is subjected to aging in operational conditions.

3.4. Application of the methodology on low-power LEDs

Section 3.2 of this chapter showed that different aging tests (radiation and thermal) were applied on GaN LEDs. We will therefore illustrate in this study the impact of aging on the optical power output, an operating parameter of LEDs. This will justify the choice of study of GaN LEDs in active storage. The main objective of this section is to justify the interest of the second and third stages of the failure analysis methodology proposed in this book. The second stage allows, from electrical and optical failure signatures, to locate degradation. It also serves as a guide for conducting physical and chemical analyses for which there will be a confirmation of the physical mechanism(s) involved in the degradation process (third stage).

3.4.1. Impact of aging on the optical power

Figure 3.15 shows the degradation of the optical power depending on aging times for GaN LEDs powered at 30 mA at $T_P = 300$ K.

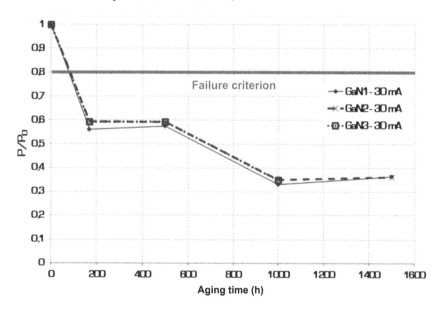

Figure 3.15. *Optical powers of three GaN LEDs before and after aging in active storage*

Note that, at 1,500 h, GaN LEDs have lost 65% of their initial optical power. In the case of GaN LEDs, it has been shown in section 3.3 of this chapter that there exists a fluorescence absorption/re-emission mechanism of the chip's coating polymer. This sensitivity to blue radiation emitted by the chip can completely vary after aging, as much as the degradation of the GaN chip that was, for example, demonstrated in 2010 by Deshayes *et al.* [DES 10]. Electrical failure and optical signatures will be useful to locate the deteriorated area(s).

3.4.2. Electrical failure signatures

It was shown in Chapter 2 that the layer close to contact (ITO) creates a Schottky barrier that hides the SRH and photonic currents. The major carriers (holes in this case) can therefore pass through the potential barrier through three tunneling mechanisms (ET, ETE and ETT). Figure 3.16 shows the band diagram of the ITO/layer p-GaN interface.

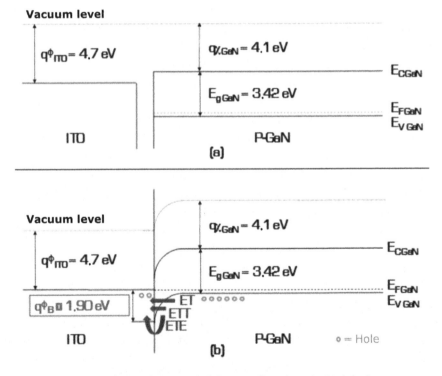

Figure 3.16. *Band diagram of the metal/semiconductor interface → ITO/p-GaN: a) off-contact layers and b) in-contact layers*

The potential barrier height is 1.903 eV before aging. The GaN LED's characteristic I(V) before and after aging is shown in Figure 3.17.

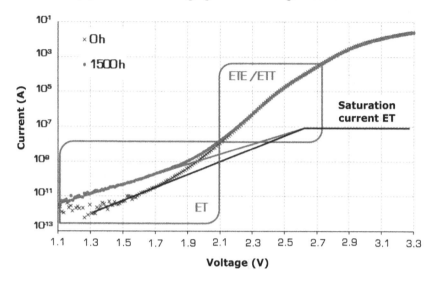

Figure 3.17. *GaN LED's characteristic I(V) before and after aging in active storage*

The characteristic I(V) of GaN LED 2 shows an increase in tunnel current of about two decades. Before aging, this current contributes only 3.51% to the total optical power because it saturates the medium to high current injection levels. After aging, its increase of two decades has resulted in a very small decrease in the tunnel barrier potential φB of 0.8%. The lowering of the potential barrier indicates that the tunnel current has increased by 15.2% and that a larger number of carriers are injected into the active region. This increase is estimated at 0.62% leading the tunnel current to contribute to 4.13% of the total optical power. The increase in the tunneling current therefore induces a very slight increase in the optical power. This phenomenon does not explain the decline in the total optical power emitted. Table 3.10 lists the values of the electrical parameters extracted from the curve I(V) before and after aging.

t (h)	R_{sh} (Ω)	φ_B (eV)	R_s (Ω)	V_{th} (V)
0	$4.5.10^{11}$	1.903	12.554	2.960
1,500	2.10^{11}	1.887	12.773	2.969

Table 3.10. *Values of electrical parameters extracted from the GaN LED 1 before and after aging*

Similarly, the series resistance has increased by only 1.7%. This small variation does not explain an optical power loss as large as 65% at 1,500 h. It is therefore considered that the variations observed by electrical failure signatures are too low, if not negligible, to explain the degradation of the optical power.

3.4.3. Optical failure signatures

Figure 3.18 shows the optical spectrum of GaN LED 2, supplied at 30 mA at T_J = 300 K, before and after 1,500 h of aging in active storage and the variation in its central wavelength and temperature depending on the aging time.

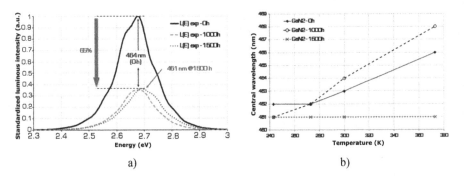

a)   b)

Figure 3.18. GaN LED 2 supplied at 30 mA at T_J = 300 K: a) optical spectra before and after aging and b) change in the central wavelength and temperature over aging time

The optical spectrum after aging indicates an optical loss of 65% and a drift of 3 nm of the central wavelength towards high energies (Figure 3.18(a)). At 1,500 h, another mechanism is added: the central wavelength remains constant in temperature. This phenomenon cannot be derived from the chip as the central wavelength always varies depending on the temperature. Indeed, it is known in the literature that the temperature varies the gap of the material constituting the component's active region [IOF 01, ROS 02]. As the temperature increases, the gap energy decreases, thereby increasing the LED's central wavelength, as confirmed by Figure 3.18(b) at 0 and 1,000 h. This means that a phenomenon external to the chip is responsible for the insensitivity of the central wavelength in temperature. We can hypothesize that the silicone oil's fluorescence was affected. If this is the case, this could mean that there has been a change in the molecular structure of the coating silicone oil and that the fluorophores are either different, or more or fewer depending on the change in fluorescence. To verify these hypotheses and explain the phenomena by localized electrical failure and optical signatures, we realized several physicochemical analyses. The latter are the subject of the next section.

3.4.4. Confirmation of failure mechanisms: physicochemical analyses

It has been seen in section 3.3 of this chapter that there is a phenomenon of fluorescence absorption/re-emission of the silicone oil coating due to the presence of SM acting as fluorophores. The fluorescence of the optical lens is discarded because the power density emitted by GaN LEDs is much too low (2.88 W.cm^{-2}) in a silicate glass. The comparison of the fluorescence emission spectra of the silicone oil excited at 360 nm before and after aging is shown in Figure 3.19.

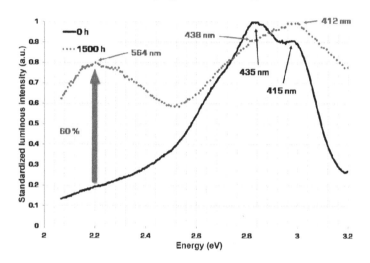

Figure 3.19. *Fluorescence spectra of a silicone oil sample before and after aging*

The fluorescence emission spectrum of the oil excited at 360 nm is interesting because it confirms a change in the silicone oil's molecular structure after aging. Indeed, from the comparison in Figure 3.19, two phenomena occur:

– A light intensity inversion of the two maximum peaks (415 and 435 nm at 0 h), which leads to one peak around 412 nm. The second peak still exists but is reduced and shifted to 438 nm. There is therefore a spectral shift of about 3 nm of the two peaks. The predominant shift is that showing a shift towards UV (415→412 nm) in accordance with the same spectral shift (464→ 461 nm) observed on the LED's optical spectrum after aging (Figure 3.18(a));

– A 60% increase in the fluorescence light intensity with a maximum at 564 nm. Here, we can assume the presence of high-molecular-weight molecules (large molecules: LM), as it was observed in the literature that LM shift the fluorescence emission spectrum towards longer wavelengths, and therefore, to lower energies [ALL 10, SWA 95, ANA 05].

The photon model of the silicone oil after aging is shown diagrammatically in Figure 3.20.

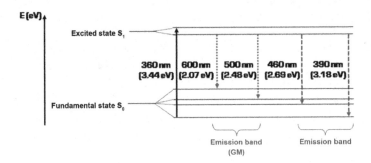

Figure 3.20. *Jablonski diagram of silicone oil excited at 360 nm after aging*

Two areas of fluorescence emission are found: 390–460 nm with a peak at 412 and 500–600 nm with a maximum at 564 nm. The second emission band (500/600 nm) appears after aging and corresponds to the appearance of LM. Table 3.11 shows the variations in the parameters extracted from fluorescence photonic models of a sample (E21) silicone oil aged in active storage.

$\lambda_{excitation}$ (nm)	0 h – E29		1,500 h – E21		FLUO Loss (%)	ABS Loss (%)
	$\lambda_{fluorescence}$ (nm)	$\eta_{fluorescence}$ (%)	$\lambda_{fluorescence}$ (nm)	$\eta_{fluorescence}$ (%)		
360	415	27.32	412	65.16	69.18	87.08 at 360 nm
	435	29.74	438	59.51	74.14	
	464	21.01	464	48.29	70.30	
	564	5.84	564	51.96	–14.94	
464	540	99.16	–	–	–	92.48 at 464 nm

Table 3.11. *Values of photonic parameters extracted from fluorescence models of the E21 silicone oil sample before and after aging*

The system lost over 90% of its absorption after aging when the latter is illuminated by the LED light. This absorption is chiefly that of the molecules that re-emit fluorescence (fluorophores). Indeed, the optical absorption of the silicone oil comprises two types of absorptions: an absorption of fluorophore molecules and absorption of molecules that are not fluorophores. In this study, only the absorption of fluorophore molecules was observed by means of fluorescence excitation spectra. In the UV, the system has gained about 15% of fluorescence in the red range (peak at 564 nm), which can be related to the presence of LM. In terms of fluorescence yield, and always under UV excitation, oil is more effective in fluorescence emission, although it lost between 69 and 70% of its initial fluorescence according to the wavelengths considered. On the other hand, an 87% decrease in the absorption of fluorophores in UV radiation shows the impact of aging on the ability of SM to absorb and retransmit fluorescence.

The fluorescence analysis therefore shows that a change has occurred in the molecular structure of silicone oil. This could be related to an effect of polymerization or cross-linking of the polymer. However, in aging, two aggravating factors that may affect a polymer are mainly present: the temperature and light. Indeed, it has already been demonstrated in the literature that one of the two factors or the combination of both can lead to polymerization or cross-linking of polymeric materials used in the assembly of optoelectronic components [RAB 95, RAB 96, HAM 00, GUI 87].

Figure 3.21 shows diagrammatically the temperature diffusion for two package temperatures: room temperature 25°C (300 K) and the aging temperature 85°C.

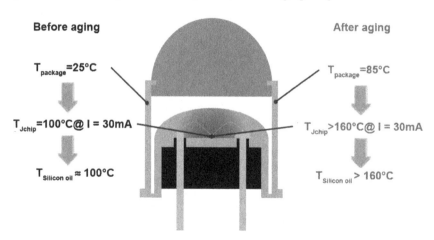

Figure 3.21. *Diagram of the temperature diffusion in the assembly of GaN LEDs*

The junction temperature given by the datasheet is estimated at about 100°C when the LED is powered at 30 mA. This means that, when the silicone oil is subjected to the aging temperature (85°C), its temperature is above 160°C. The first hypothesis resides in the principle of modifying the molecular structure of the polymer activated by the temperature. This change may result in a polymerization or cross-linking mechanism of the silicone oil.

To understand the role of the temperature and verify our hypothesis, we conducted a differential scanning calorimetry analysis (DSC). Figure 3.22 shows the DSC spectrum of the oil before and after aging.

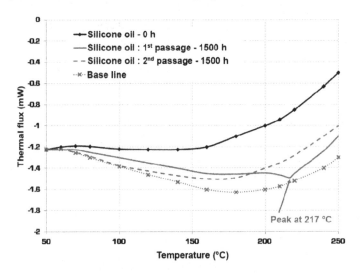

Figure 3.22. *DSC spectrum of a silicone oil sample before and after aging*

Before aging, no fusion or decomposition process has appeared. However, at 1,500 h, a large peak appears at 217°C on the first passage. Its size confirms the nature of the analyzed material (polymer) and the disappearance of the peak in the second passage confirms that there is a process of change in the molecular structure of the oil after aging. This suggests that a polymerization or cross-linking mechanism is involved in the degradation of fluorescence inducing a major loss of optical power.

To confirm this mechanism, we conducted mass spectroscopy and RMN ^{1}H analyses. Figure 3.23 compares the mass spectra of oil samples before and after aging.

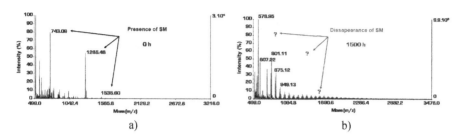

a)   b)

Figure 3.23. *Mass spectra of silicone oil samples: a) sample*
E29 before aging and b) sample E21 after aging

Note the disappearance of the three peaks (743, 1,285 and 1,535) after aging. These correspond to the SM.

Figure 3.24 provides a comparison of RMN ^{1}H spectra of different silicone oil samples:

– E29 and E30 are un-aged samples;

– E24 and E26 were subjected to aging in passive storage (1,500 h/85°C);

– E21, E22 and E23 underwent similar aging in active storage as the LEDs.

This choice was made in order to discriminate the impact of temperature from that of the light on the modification of the molecular structure of the oil and thus the variation in the oil's fluorescence emission.

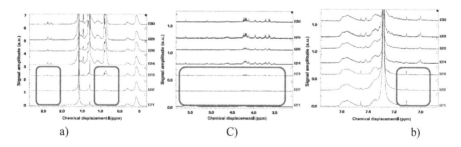

a)   C)   b)

Figure 3.24. *RMN ^{1}H spectrum of several silicone oil samples before and*
after aging: a) aliphatic region (methyl), b) intermediate region (siloxane)
and c) aromatic region (phenyl)

The boxed areas are the impacted areas (disappearance of peaks). RMN spectra of the silicone oil confirm the results of the mass spectrometry. Indeed, for the samples aged in active storage, the peaks corresponding to SM disappear in the three regions (aliphatic, aromatic and intermediate). When observing the diffusion spectrum (dosimetry) of the most impacted sample (E21: active storage), other information appears. Its dosimetry spectrum is shown in Figure 3.25 before and after aging.

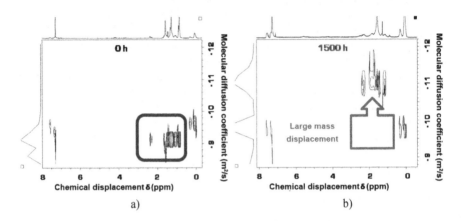

Figure 3.25. *Dosimetry spectra of two silicone oil samples: a) E29 un-aged and b) E21 after aging in active storage*

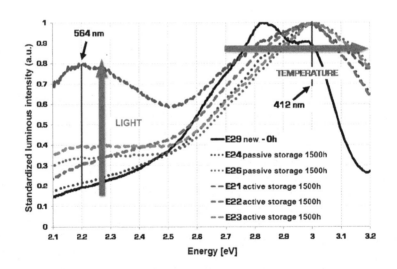

Figure 3.26. *The impact of the temperature and light from the LED on the coating silicone oil before and after aging*

The presence of LM (mass relative to SM by a factor of 100) is confirmed after aging. This may be in agreement with a mechanism of polymerization or cross-linking of the silicone oil as it showed a mass change towards large molecular weights.

Finally, to discriminate the effect of the temperature from that of the light, a complementary fluorescence analysis was performed on the various samples aged with or without light. Figure 3.26 compares the fluorescence emission spectra of the silicone oil samples E29, E21, E22, E23, E24 and E26 excited at 360 nm.

The standardization of these spectra allows the calculation of the ratio between the two maxima of each area (red/blue). Table 3.12 shows the results of this calculation by adding the fluorescence yield at 564 nm and the absorption loss at 360 and 464 nm.

t (h)	Parameter	New	Passive storage		Active storage		
		E29	E24	E26	E21	E22	E23
0	Ratio (%)	80	–	–	–	–	–
1,500		–	78	68	20	70	60
0	$\eta_{fluorescence}$	5.84	–	–	–	–	–
1,500	at 564 nm (%)	–	51.70	49.68	51.96	52.29	49.07
1,500	ABS losses at 360 nm (%)	–	83.49	74.77	87.08	83.70	76.20
	ABS losses at 464 nm (%)	–	79.60	77.64	92.48	85.44	79.05

Table 3.12. *Values of photonic parameters of the different samples before and after aging*

Generally, the red/blue ratio is lower when light and temperature are combined than when only the temperature represents an aggravating factor. This agrees with the fact that the fluorescence yield is higher in the case of samples aged in active storage than that of samples aged in passive storage.

To verify the impact of the light only on the oil's fluorescence, a silicone oil sample was illuminated at 464 nm (the LED light) for 16 h. Figure 3.27 shows the fluorescence emission spectrum for an excitation wavelength of 360 nm.

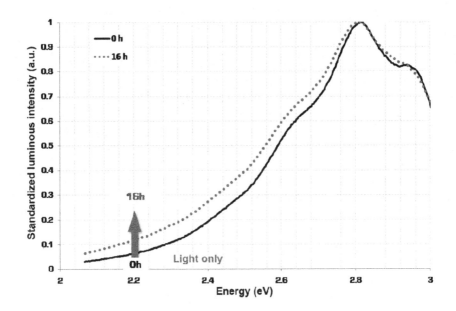

Figure 3.27. *Impact of the LED light (464 nm) on the coating silicone oil after 16 h light at 464 nm*

This experiment validates the impact of the LED light on the fluorescence emission: the light is partly responsible for the increase in fluorescence in the red range (564 nm), which can be linked to the occurrence of LM, while the temperature plays a role in the spectral shift in the shorter wavelengths (high energies).

Electro-optical failure signatures have therefore made it possible to locate the failure at the silicone oil's level which, under the LED's radiation, fluoresces. Through physicochemical analyses, the failure mechanism explaining the optical power loss was highlighted: the modification of the molecular structure of the silicone oil activated photo-thermally induces a loss of fluorescence greater than 69% due to a high reduction in absorption of the LED light (464 nm) by the fluorophores (PM and GM), estimated at 90%. This change could be related to a mechanism of polymerization or cross-linking of the oil after aging.

3.5. Summary of results and conclusions

This section has allowed highlighting the sensitive areas of low-power (30 mW) GaN LEDs when subjected to aging respecting operational conditions (active storage: 1,500 h/85°C/I_{rated}).

Indeed, the latter consists of a polymeric material (PMPS/PDMS copolymer), which encapsulates the chips. The aim was twofold:

– to show the impact of aging on the polymer. In fact, the latter is transparent to IR radiation, but strongly interacts when it is subjected to shorter wavelengths (blue/UV LED →GaN);

– to justify the contribution of physicochemical analyses in the applied failure analysis methodology.

We thus showed, through the methodology proposed in this study, how to establish a failure analysis on optoelectronic components using the electro-optical signatures and, where necessary, physicochemical analyses.

GaN LEDs have undergone an optical power loss of 65% for all the LEDs at 1,500 h. To determine the failure mechanism explaining such optical losses, we have relied on the extraction of electro-optical failure signatures, allowing us to locate the failure. Electrical failure signatures were estimated to be too low to explain the optical power degradation. However, analysis of the optical spectrum confirmed an optical loss of 65% and showed a drift of the central wavelength of 3 nm. Secondly, the central wavelength temperature remained constant at 1,500 h. Under the conditions of the imposed measurements, only an external phenomenon to the chip may be responsible for this thermal insensitivity. This allowed us to assume that the fluorescence of the silicone oil was affected, which could mean a change in its molecular structure with either different fluorophores or a larger or smaller number of fluorophores according to the change in fluorescence. To verify all of these hypotheses and explain the localized phenomena by electrical failure and optical signatures, we conducted several physicochemical analyses. The fluorescence analysis of silicone oil sample (E21) excited in the UV range (360 nm) showed two main phenomena:

– A light intensity inversion of the two maximum peaks (415 and 435 nm before aging) with a red-shift of about 3 nm for both peaks. The predominant shift towards UV (415→412 nm) in accordance with the same spectral shift (464→461 nm) is observed on the LED's optical spectrum after aging;

– A 60% increase in the fluorescence light intensity with a maximum at 564 nm. Here, we have assumed the presence of high-molecular-weight molecules (large molecules: LM).

We have also determined a higher absorption loss of 90% at 1,500 h when the silicone oil is illuminated by the light from the LED (464 m). The optical absorption of the silicone oil comprises two types of absorptions: absorption of the fluorophore molecules (re-emitting in fluorescence) and absorption of molecules that are not fluorophores. In this study, only the absorption of fluorophore molecules was observed by means of fluorescence excitation spectra. In the UV, the system has gained about 15% of fluorescence in the red range (peak at 564 nm), which can be justified by the presence of LM playing the role of new fluorophores. This has oriented our study to the possibility of a change occurring in the molecular structure of the silicone oil, which could be related to a polymerization or cross-linking mechanism. However, during aging, two aggravating factors that may affect a polymer are mainly present: temperature and light. After having shown that, during aging, oil temperature exceeded 160°C, we wanted to confirm the role of the latter by performing a DSC analysis. The results were used to validate this process of modification of the molecular structure of the oil by indicating a DSC peak at 217°C. To confirm for a second time the mechanism of polymerization, mass spectral analyses and ¹H RMN were performed. The mass and RMN spectra showed that there has been a disappearance of SM (former fluorophores) after aging. In addition, the RMN spectrum dosimetry proved the existence of LM (new fluorophores) at 1,500 h. These phenomena could be consistent with a mechanism of polymerization or cross-linking of the silicone oil, as it showed a mass change tending towards large molecular masses. Finally, to discriminate the effect of the temperature from that of the light, a complementary fluorescence analysis was performed on the various aged samples with or without light. This has helped to highlight that light is partly responsible for the increase in fluorescence in the red (564 nm) that can be linked to the development of LM, while the temperature plays a role in the red-shift in the shorter wavelengths.

A summary diagram is shown in Figure 3.28.

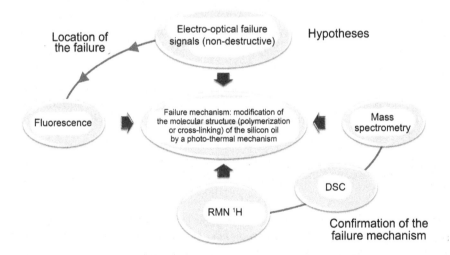

Figure 3.28. *Summary of the GaN LED results*

Electrical and optical failure signatures allow, using hypotheses, to locate the fault, which is connected to the fluorescence of the silicone oil, while the physicochemical analyses such as RMN, DSC and spectrometry ground have allowed us to confirm the failure mechanism: modification of the silicone oil's molecular structure activated photo-thermally and inducing a loss of fluorescence greater than 69% due to a strong decrease in the absorption of the LED light (464 nm) by the fluorophores (SM and LM), estimated at 90%. This change could be related to a mechanism of oil polymerization or cross-linking. In this section, the beneficial contribution of physicochemical analysis was emphasized.

It will be shown in the last chapter (Chapter 4) that this methodology can be applied to industrial issues in public lighting. The objective will be to explain degradation mechanisms in yellowing of the white light of white power LEDs.

4

Integration of the Methodology
Starting from Component Design

Power light-emitting diodes (Leds) for public lighting are distinguished from other Leds by the need to obtain a color yield index (CYI) greater than 85, the highest possible optical power ($>>$ 1 W), high luminous efficiency ($>>$ 100 lm.W^{-1}), excellent thermal management with thermal resistances of assemblies lower than 4°C.W^{-1}, and the lowest possible manufacturing cost ($<\$$ 4/lamp).

We saw in Chapter 3 that the two aggravating factors for low-power Leds are temperature and light with a wavelength below 500 nm. These two factors can lead to a significant optical power loss ($>$ 60%) and a spectral shift involving a color change. The coating silicone oil's molecular structure changing mechanism activated by the photo-thermal effect was thus demonstrated. The latter could be linked to a polymerization or a cross-linking mechanism of the Si oil.

In the case of public lighting, white power Leds are fully subject to these thermal and photonic constraints. This phenomenon is aggravated by the fact that the junction temperature of the power Leds is considerably greater than that of the Leds discussed in Chapter 3.

The aim of this chapter is to apply the methodology of failure analysis, operated and presented in the previous chapter, on power Leds manufactured with a complex polymer coating. To understand the different failure mechanisms, we established, in collaboration with an LED assembler, a process by component batch. Thus, we implemented a series of Leds with a polymer coating and a single series with a polymer charged in YAG:Ce phosphor.

These components will be subjected to aging in active storage (85°C/550 mA/500 h), and the challenge will be to determine the optimal structure to improve the studied white power LEDs. We will therefore present the context of this study, and then we will detail the specifications and the aging campaign that will be carried out. We apply the same architectural methodology as the one developed in Chapter 3.

The purpose of this study is twofold:

– to demonstrate that the failure analysis methodology developed in this research can also be integrated starting from a component's design, and is adaptable to high powers;

– to provide technological solutions responding to an identified need: to explain the physical failure mechanisms leading to yellowing of the white light of power LEDs.

The key point of this chapter is to show that failure physics allows determining the sensitive area as well as the degradation process. These elements will be used to improve the assembly and thus to make it efficient.

4.1. Mission profile for public lighting

The success of the market of GaN power technologies for public lighting has involved considerable development since the early 2000s. Europe, being the leader in this field, has heavily invested in the replacement of incandescent and fluorescent tubes with LED lamps. Today, the dominant technology for achieving a power white light emission (> 1 W) is that of GaN LEDs associated with a phosphor coating.

Applications related to public lighting, particularly of cities and highways, have a specific mission profile. We will concisely present the different environmental constraints induced by this type of application. The second aspect is the heat flow that becomes very critical at powers greater than 1 W. Various designs have been proposed and the objective of the IMS Laboratory is to assist in the technological choice with the aid of physical failure analysis.

Photonic aspects coupled with thermal ones, especially in the assembly, are discussed in this chapter and are one of the skills acquired during this research. The theoretical skills developed in Chapter 2 will be strongly solicited to successfully carry out this industrial collaboration.

4.1.1. Context and project objectives

In France, two major players represent the keystone in terms of research and development of white power LEDs for public lighting:

– The Center of Research on Heteroepitaxy and its Applications (CRHEA) of Sophia Antipolis of the University of Nice. CRHEA is a laboratory of the National Center for Scientific Research (CNRS). The high specificity of this research laboratory is the mastery of heteroepitaxy and therefore its skills are centered on the "bare chip". The IMS laboratory strongly collaborated with CRHEA in the early 2000s with L. Hirsch [HIR 01, HIR 02]. The heart of the CRHEA laboratory's activities is the growth of materials by epitaxy: thick, thin film, quantum heterostructures (wells, wires and casings) or nanostructures. These materials are part of the large bandgap semiconductor theme: gallium nitride (GaN, InN, AlN and alloys), zinc oxide (ZnO) and silicon carbide (SiC) [CRH 16] ;

– The Laboratory of Electronics and Information Technology (LETI), division of the Commission for Atomic and Alternative Energies (CEA). This research center is particularly expert in the assembly of LEDs: phosphor report and coating. CEA-LETI is involved in a wide range of research and development programs whose purpose is improving the performance, quality and reliability of the LED technology. It has worked in the emerging field of LED lighting since 2006. Its initial research and development programs were limited to the upstream of the industrial value chain, primarily focused on new semiconductor materials, such as zinc oxide, and breakthrough technologies such as LED nanowires. Since then, more new "downstream" projects were launched on key technologies relative to thermal management, light extraction or conversion of wavelengths for example [CEA 16]. The problems in terms of reliability of CEA-LETI are in the process of manufacturing of GaN power LEDs. The implementation of LED chips in packages and in a complete system may be detrimental from the standpoint of the device's lifetime. The detailed study of failure causes involving the chip's internal structure and the interaction of the assembled LED with the environment represents a challenge whose interest intersects with our methodology.

In this context, we realized an aid to the design by using physical failure analysis. Indeed, the behavior of aging LEDs must be known, and physical phenomena explaining the failure included for finding the changes to be made to the design in order to improve reliability.

In this way, we help to meet the needs identified at the industrial level. Several research and development projects have emerged on this topic in collaboration with Philips Lighting [BER 11, MAR 11].

Our study's main objective is to identify and explain the failure mechanisms of the shift to the yellow of the light emitted by YAG:Ce power LED technology under environmental constraints related to public lighting (current, temperature). The purpose of this study is twofold:

– to understand the physical failure phenomena in order to identify the technology's sensitive areas;

– to find technological solutions in order to reduce the sensitivity of fragile areas and therefore to increase lifetime. This process makes the technology robust, and therefore, controls its quality and, more broadly, its reliability.

4.1.2. Environmental requirements and constraints in public lighting

Public lighting is classified into two categories: indoor and outdoor. The requirements for this field in 2011 in terms of luminous efficiency (lm.W^{-1}), luminance (lm), color rendering index (CRI) and lifetime are summarized in Table 4.1 [SPE 11]. The CRI was established to indicate how colors appear under different light sources. A mathematical comparison is generally used to determine how a light source moves eight colors defined on the color space (CIE), compared with the same colors lighted by a reference source of the same color temperature. If there is no difference in appearance, the light source has a CRI of 100 by definition. From 2,000 to 5,000 K, the reference source is the black body and, over 5,000 K, it is a form of well-defined daylight [COL 07].

Type of LED lighting	Luminous efficiency (lm.W^{-1})	Luminance	CRI	Lifetime (h)
Indoor	20–50	200–750	75–80	25,000
Outdoor	29–80	50–300	50–80	35,000

Table 4.1. *Pubic lighting requirements*

The color temperature of a light source is defined from the absolute temperature in Kelvin (K) from the radiation of a heated black body (from 2,000 to 10,000 K), with a similar emission spectrum to that of the light source. Sources with low color temperatures (<4,000 K) have a color tending towards yellow and red, known as hot color. Sources with high color temperatures (> 4,000 K) have a color tending towards blue, known as cold color. The sun, being at 5,750 K, corresponds to a white light with a little yellow. The white colors accepted for indoor and outdoor lighting are summarized in Table 4.2 [ENE 08].

White nominal temperature (K)	Temperature drifts and accepted tolerances (K)
2,700	2,725 ± 145
3,000	3,045 ± 175
3,500	3,465 ± 265
4,000	3,985 ± 275
4,500	4,503 ± 243
5,000	5,028 ± 283
5,700	5,665 ± 355
7,000	6,530 ± 510

Table 4.2. *White Temperatures accepted in indoor/outdoor lighting and tolerated drifts*

Figure 4.1 shows the correspondence of white temperatures used for power LEDs in public lighting.

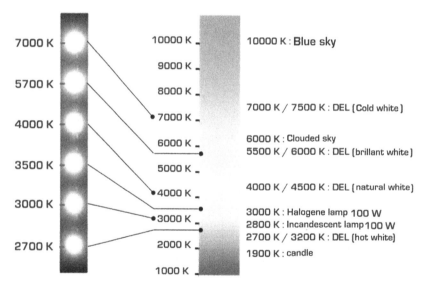

Figure 4.1. *a) Main white temperatures (LED lighting) and b) white temperature range for public lighting. For a color version of the figure, see www.iste.co.uk/deshayes/reliability2.zip*

The LED lamp failure criterion is 30% loss of luminance at the end of life [ENE 08]. In 2010, LEDs fulfilled these requirements. Table 4.3 shows the roadmap of the luminous efficiency for two white fields in public lighting from 2010 to 2020 [BAR 11].

Luminous efficiency (lm.W^{-1})	2010	2012	2015	2020
Hot white (2,780 – 3,710 K)	96	141	202	253
Cold white (4,746 – 7,040 K)	134	176	224	258

Table 4.3. *Roadmap of the luminous efficiency from 2010 to 2020 according to different white light sources*

We note that the luminous efficiency is increasing every year. This is a factor related to the component's operating temperature. Indeed, for the same supply current (typically 350 mA), optical power increases every year. This means that the current is an aggravating factor related to public lighting requirements.

In France, the main environmental constraint associated with public lighting is ambient temperature. In some countries such as Asian countries, humidity is a constraint that is as important as temperature. The reliability standards require operating temperature ranges from –40°C to +60°C, +85°C or 100°C for LED lamps [CRE, LED].

The two aggravating factors selected for this study are thus the current and temperature.

4.1.3. *Studied technologies*

To study aging in operational conditions of white power LEDs, EZ1000-type LEDs manufactured by CREE were selected and assembled on a MCPCB (metal PCB) STAR aluminum base. The electrical and optical parameters of the latter are referenced in Table 4.4.

Parameter	Typical value at 300 K and I = 350 mA
$I_{nominal}$	350 mA
$\lambda_{central\ GaN\ LED}$	451 nm
$P_{optical}$	250 mW
Operational temperature range (°C)	–40°C to 100°C
$\Theta_{emission}$	115 – 130 °
T_{white}	4,300 K
Flux of LED with phosphor	55 lm
Flux of LED without phosphor	10 lm
$V_{nominal}$	3.3 V

Table 4.4. *Electrical and optical parameters of CREE EZ1000 LEDs*

A total of 10 LEDs were assembled and supplied:

– two sample CREE XLamp XP-E-type Leds;

– four CREE EZ1000-type blue Leds on MCPCB STAR base;

– four CREE EZ1000-type white Leds on MCPCB STAR base.

The ceramic support, wherein the GaN chips are located, has been welded to the $In_{50}Sn_{50}$ on their MCPCB STAR base, and deposited and then sedimented the coating to the YAG:Ce phosphor.

4.2. Aging campaign and component description

4.2.1. Aging campaign specifications

Given the aggravating factors (current and temperature), we chose to apply aging in active storage for 500 h. The temperature was chosen based on qualification standards for public lighting, 85°C. The supply current has been determined from the linearity zone (400–700 mA) of the characteristic of the junction temperature as a function of the supply current, 550 mA. Table 4.5 summarizes all the aging implemented on eight LEDs.

LED technology	Aging temperature (°C)	Supply current (mA)	Aging duration (h)	Number of aged LEDs
Blue LEDs	85	550	500	4
White LEDs	85	550	500	4

Table 4.5. *Aging campaign in active storage*

The electro-optical and thermal characterizations were performed at 0, 96, 200 and 500 h:

– electrical characteristics I(V) at junction temperature $T_J = 300$ K;

– thermal characteristics $T_J(I)$ at ambient temperature $T_P = 300$ K for I varying from 100 to 700 mA in 100 mA steps including I_{th} thermal threshold current;

– spectral characteristics L(E) at $T_J = 300$ K for the currents I = 200 mA, I_{th}, 400 mA and 700 mA;

– spectral characteristics in power L (E) at $T_P = 300$ K for a current I = 350 mA;

– luminous flux at 350 mA and $T_P = 300$ K;

– MacAdam's ellipse at 350 mA and T_P = 300 K. Each ellipse represents the smallest perceptible difference between two close colors;

– color: measure of the temperature of white at 350 mA and T_P = 300 K;

– measurement of voltage at 350 mA and T_P = 300 K.

4.2.2. Technological description of LEDs

All information relating to the chip and its assembly's structure come from the technical documentation of different manufacturers. Additional bibliographic research on the chip and its assembly has confirmed this information.

4.2.2.1. MQW InGaN/GaN structure power LEDs

The GaN LED is an EZ1000 chip manufactured by CREE. The structure of the latter is given in Figure 4.2.

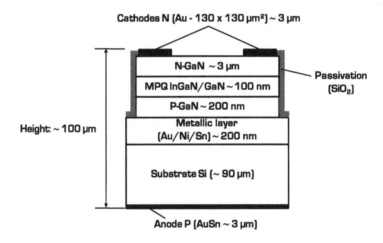

Figure 4.2. *Cross-section of CREE's EZ1000 LED*

This LED is a vertically emitting structure. It has a lateral passivation layer (SiO_2), which reduces the leakage current escaping from the chip's edges. The superior N contact is composed of a grid on the chip's surface and of two square pads (130×130 μm²), with a thickness of 3 μm, in Gold (Au) on which the bonding wires are connected. The chip's rear face (anode P) is an $Au_{80}Sn_{20}$ alloy of the same surface area as the chip (980×980 μm²) of about 3 μm thickness, deposited on the Si substrate.

To develop this structure, a SiC substrate is generally used to grow by MOCVD the layers of the active region from the layer N-GaN to the P-GaN layer [HON 08]. Thus, an un-doped GaN layer with a thickness of 2 μm was deposited on a SiC substrate. On the latter, an N-doped GaN layer ($Si/10^{17}-10^{19}$ cm^{-3}) 3 μm in thickness was deposited to grow the MQW InGaN/GaN structure, of 100 nm thickness, covered by a P-doped GaN layer ($Mg/10^{18}$ cm^{-3}) of 200 nm thickness. A metal layer (multilayer Au/Ni/Sn), 200 nm in thickness, was deposited by pulverization on the P-GaN layer. This layer plays the role of assembly on which a Si substrate is deposited. It also has a reflective silver layer, located between its surface and the P-GaN layer, to prevent significant absorption of light by the silicon substrate (Si). The vertical emission of the LED is therefore due to this metal layer. Finally, a UV laser (193 nm ArFexcimer or 355 nm Nd: YAG laser) is generally used to separate the SiC substrate of the undoped GaN layer (laser lift off –LLO technique). The undoped GaN can be removed by plasma etching ICP-RIE (inductively coupled plasma-reactive ion etching) using BCl_3/Cl_2 to expose the GaN layer [CHO 08].

4.2.2.2. Structure and assembly processes

The assembly of CREE EZ1000 LEDs is shown schematically in Figure 4.3. Its structure is the same for both types of LEDs apart from the silicone coating.

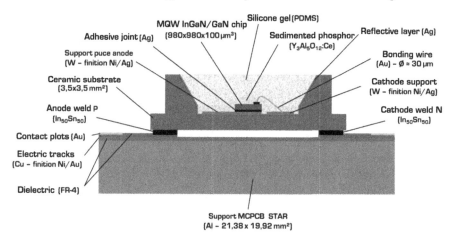

Figure 4.3. *Cross-sectional view of the CREE EZ1000 LED assembly on STAR base*

Light is emitted by the EZ1000 chip bonded by an adhesive joint (Ag: 85%) on its support (W) with coating Ni (2.5 nm)/Ag (200 nm) with a reflectance of around 95% in order to improve light extraction [CHE 06]. The carrier (W) is deposited on a ceramic substrate of alumina (Al2O3) of surface 3.5 × 3.5 mm² on which is deposited a reflective layer (Ag) of approximately 95% reflectance [KIM 05b]. A bonding wire with a diameter of 30 μm connects each pad (130 × 130 μm²) of the cathode (N side) to the support W deposited on the ceramic substrate. This technology of realization of ceramic casings is known as the HTCC (high-temperature co-fired ceramic).

The rear face of the ceramic substrate has two rectangular copper pads (1 × 3.5 mm²) corresponding to the chip's anode and cathode. Both sides of the ceramic substrate are connected by external vias. The ceramic substrate is soldered at 150°C to the $In_{50}Sn_{50}$ on the MCPCB STAR support (21.38 × 19.92 mm²); 1.5 mm thick aluminum (metal). A dielectric (FR-4: Type 4 flame retardant) 100 μm thick, is present over the entire surface of the STAR MCPCB. It is generally used in electronics for delaying ignition of circuits printed on PCBs. This material is mainly composed of fiberglass reinforced epoxy resin to increase its glass transition temperature (120–180°C depending on the epoxy resin). In the case of GaN power LEDs, the FR-4 helps not only to prevent ignition of the PCB during component operation and to electrically isolate the aluminum PCB from Cu tracks and Au contact pads (LED power), but also to allow the adhesion of the copper and aluminum [DEM 07]. The finishing of the Cu tracks is achieved by deposition of nickel (a few μm), which serves as an inter-metallic diffusion barrier on which a layer of gold (<1 μm) is deposited. Indeed, gold plating directly on the Cu tracks is not recommended because of the risk of inter-diffusion between copper and gold [GFI 09].

The chip's coating is a silicone gel NUSIL manufactured by Lightspan with a refractive index of 1.55 – 411 nm at 25°C. This is a polydimethyl siloxane polymer (PDMS). This gel's operating temperature range is from −40°C to 200°C. For the four white LEDs, the coating is a mixture of PDMS silicone gel (85%) and phosphor powder (15%) manufactured by Phosphor Technology. The phosphor of type YAG:Ce with a base of cerium (Ce^{3+}) doped yttrium aluminum ($Y_3Al_5O_{12}$) garnets, is then sedimented in a uniform layer around the chip and bottom of the ceramic casing. This allows converting the blue light emitted by the chip into white light with a CRI generally lower than 75 [OH 10].

4.3. Physical failure analysis

One of the main objectives of this section is to rely on the failure analysis methodology discussed in Chapter 3 to identify failure mechanisms inducing degradation of blue and white power-assembled LEDs. This approach helps to show that the methodology can be used to help design the assembly of a component and can lead to technological solutions in order to improve performance and reliability of white power LEDs used for public lighting. The first section identifies the electrical, optical and thermal failure signatures for both technologies. This will allow pre-locating degraded areas. The use of physicochemical analyses of fluorescence and X-ray diffraction types will confirm the degradation mechanisms occurring in the degraded areas. This will help both to explain the physical origin of the degradation due to aging in active storage, and locate areas susceptible to change to increase the assembly's robustness.

4.3.1. Location of degraded areas: electro-optical and thermal failure signature

Figure 4.4 shows the evolution of the relative optical power (P/P$_0$) as a function of aging time for two LEDs with and without YAG:Ce phosphor.

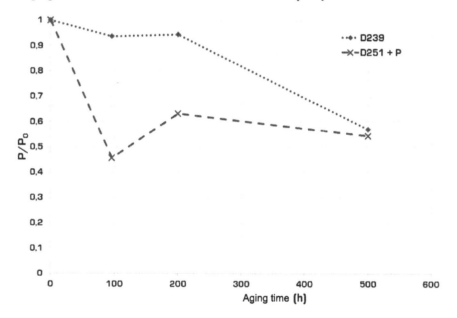

Figure 4.4. *Evolution of the relative power P/P$_0$ as a function of aging time for I = 350 mA @ TJ = 300 K/D239: LED without phosphor (blue); D251: LED with phosphor (white). For a color version of the figure, see www.iste.co.uk/deshayes/reliability2.zip*

We observe an optical power loss of about 40% at 500 h for both types of LEDs (with and without phosphor). Such a rapid decrease in the optical power shows a behavior that is quite rare for white power LEDs. We remind the reader that the results demonstrated in this book concern only the studied blue and white LEDs. The second observation is that the degradation kinetics depends on the LED type. The blue LED has exceeded 10% degradation only after 100 h, whereas the white

LED degraded by more than 50% after 96 h. To verify this phenomenon, it is interesting to look at the optical power losses for all LEDs. Table 4.6 summarizes the total loss of optical power for all analyzed LEDs.

Aging time (h)	Losses of LEDs without phosphor (%)				Losses of LEDs with phosphor (%)		
	D238	D239	D240	D241	D251	D252	D254
0	0	0	0	0	0	0	0
96	−4.74	6.32	1.15	−5.60	54.39	−7.75	−3.36
200	−0.40	5.53	1.54	3.45	36.82	7.04	8.72
500	35.97	43.08	34.62	20.26	45.68	40.49	24.16

Table 4.6. Total losses of optical power ($\Delta P/P_0$) during aging in active storage

We observe the same degradation kinetics between the blue and white LEDs. An observation of colorimetry of white LEDs allows us to identify the impact of aging on the temperature of the white color. Table 4.7 shows the values of the white color temperatures for the three LEDs with phosphor between 0 and 500 h.

Aging time (h)	White temperatures (K)			Drift (%)		
	D251	D252	D254	D251	D252	D254
0	4,316	4,302	4,327	0	0	0
96	4,572	4,399	4,422	−5.93	−2.25	−2.20
200	4,211	4,354	4,373	2.43	−1.21	−1.06
500	4,159	4,181	4,282	3.64	2.81	1.04

Table 4.7. White temperature drifts for LEDs with phosphor

All color equivalent temperatures decreased (from 1 to 3.6%) at 500 h. This means, according to Figure 4.1, after aging, the white drifts towards yellow.

After the LEDs' performance analysis phase, we shall focus on the location of the degradation and the physical analysis of degradation. The first step of this study is to analyze electrical failure signatures. The latter is mainly connected to the bare chip at a very low current level. Figure 4.5 shows the I(V) characteristics of LEDs 239 and 251 before and after aging.

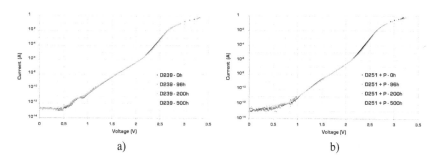

a)   b)

Figure 4.5. *I(V) characteristics at 0 and 500 h: a) blue LED 239, b) white LED 251*

The electrical characteristics of all the LEDs do not show any variation in current for the blue LEDs, and a non-significant change (<1%) in the thermionic current (ETT), which acts only on 5% of the optical power. This behavior is identical for all 3 white LEDs. Therefore, these variations cannot be connected to the optical losses of over 40%. To complete the electrical study, we are now interested in the optical and thermal failure signatures.

The optical behavior differs depending on the type of LEDs. It is therefore necessary to separate the studies of blue LEDs from those of white LEDs.

4.3.1.1. Optical and thermal signatures for LEDs without phosphor

Figure 4.6 shows the optical spectra of the blue LEDs powered at 350 mA at $T_p = 300$ K before and after aging and recalling the loss of optical power for each spectrum.

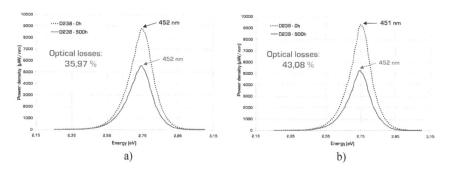

a)   b)

Figure 4.6. *Optical spectra L (E) of the blue LEDs powered at 350 mA at 0 and 500 h: a) LED 238 and b) LED 239*

We find the optical power losses of Table 4.6 with a shift of the central wavelength of about 1 nm for all LEDs. Table 4.8 shows this shift at T_P = 300 K, and the wavelengths at T_J = 300 K before and after aging.

Conditions	Aging time (h)	Central wavelength (nm)			
		D238	D239	D240	D241
T_P = 300 K, I = 350 mA	0	452	451	451	451
	500	452	452	452	452
T_J = 300 K,	0	445	443	444	444
I = 700 mA	500	445	443	444	444

Table 4.8. *Shift of central wavelength for LED without phosphor at T_P and T_J = 300 K before and after aging*

At T_J = 300 K and I = 700 mA, no variation is observable. This indicates that the effective current in the active zone remained constant after aging. At T_P = 300 K and I = 350 mA, there is a spectral shift of 1 nm. The latter may be due to a drift of the junction temperature leading to a modification of the heat flow within the active region since the gap of the latter is dependent on temperature [CHO 99]. Figure 4.7 shows the variation in the junction temperature T_J, in current, before and after aging at an ambient temperature T_P of 300 K.

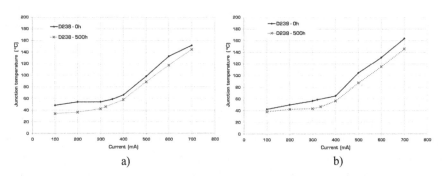

a)   b)

Figure 4.7. *Variation in T_J in current at T_P = 300 K before and after aging for the blue LEDs: a) LED 238 and b) LED 239*

This temperature characteristic has two parts: a limit of 100–300 mA for which the temperature stabilizes at about 45°C, and a linear part of 400–700 mA with a slope of $0.31°C.mA^{-1}$. This allows extracting the assimilable current at a thermal threshold current close to 350 mA, and the associated junction temperature of about 60°C. The characteristic T_J (I) shows an overall decrease (about 10°C) in the LEDs' junction temperature. This means that the thermal management of LEDs is improved after aging. Table 4.9 summarizes the changes in thermal parameters extracted from the characteristic T_J (I) for a current of 350 mA at T_P = 300 K before and after aging.

LED	T_J (°C) – 0 h	T_J (°C) – 500 h	ΔT_J (°C)	ΔR_{th} $(K.W^{-1})$	Gradual change of slope (%)
D238	59.49	50.09	−9.40	−14.56	−1.06
D239	60.88	49.56	−11.31	−16.64	−8.10
D240	67.07	56.79	−10.28	−16.63	−17.62
D241	57.58	57.06	−0.51	−2.78	−11.99

Table 4.9. *Variations in the thermal parameters of the blue LEDs at 350 mA and at the linear slope portion (400–700 mA) of the T_J(I) curve at T_P = 300 K before and after aging*

R_{th}: the LED's thermal resistance

The decrease in the LED's thermal resistance confirms better thermal management. Now, we have seen that the LEDs' central wavelength λ_C increases by 1 nm at T_P = 300 K after aging for most LEDs. In general, the gap of the active region increases (in energy) when the temperature decreases. This implies that λ_C must decrease when T_J decreases, which is not the case here. This observation indicates that the temperature dependence of the center wavelength has been modified. To confirm this phenomenon, we have estimated, from the literature, the slope λ_C as a function of T_J ($d\lambda_C/DT_J$) before aging. Indeed, this variation is similar to the gap of the active area, and was estimated at 0.034 nm.K^{-1}. Considering the spectral shift of each LED, the value of this parameter was estimated at 500 h. Figure 4.8 shows the variation of λ_C as a function of T_J for the blue LED 239 before and after aging.

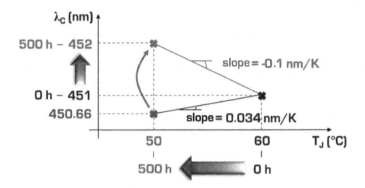

Figure 4.8. *Variation in λ_C as a function of T_J before and after aging for the blue LED 239*

At 500 h, with approximately 10°C of T_J loss and a slope of 0.034 nm.K^{-1} at the initial state, the LEDs' central wavelength λ_C should have been less than 451 nm. However, after aging, we have seen it increased by 1 nm. The slope $d\lambda_C/dT_J$ became negative. Table 4.10 shows the impact of aging on the dependence on temperature of the central wavelength of the LEDs without phosphor.

LED	$d\lambda_C/dT_J - 0$ h	$d\lambda_C/dT_J - 500$ h	Variation (%)
D238	0.034	0.034	0.00
D239	0.034	− 0.1	61.53
D240	0.034	− 0.1	65.06
D241	0.034	− 1.957	98.26

Table 4.10. *Impact of aging on the slope dλC/DTJ for LEDs without phosphor*

These results highlight a thermal failure signature defined by a variation in the slope of the central wavelength of the LEDs in temperature after aging. This confirms the modification of the thermal flow of the LEDs at 500 h, which results in a spectral shift of 1 nm and a reduction in junction temperature of about 10°C at 350 mA and $T_P = 300$ K.

The device is therefore improved from a thermal standpoint, which should thus result in improved LED performance. Despite this, the optical power drops are consistent and it is therefore likely that the polymer coating has deteriorated. A physicochemical analysis of fluorescence would confirm this hypothesis. This will be the subject of section 4.3.2.

4.3.1.2. *Optical and thermal signatures of LEDs with phosphor*

The addition of the phosphor in the coating oil has several effects. The first is the conversion of blue light into white light. This phenomenon was explained by Tomiki *et al.* emphasizing the role of Ce^{3+} in an $Y_3Al_5O_{12}$ (YAG) host matrix. Figure 4.9 diagrammatically shows the energy band as a function of the coordinate of the R configuration of the Ce^{3+}, the associated emission spectra and Jablonski diagram of the Ce^{3+} atom in a YAG crystal [TOM 91].

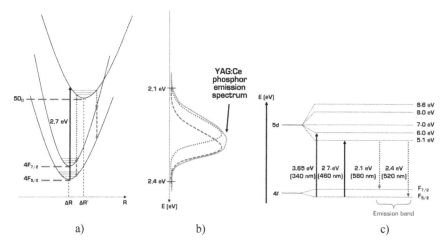

a)   b)   c)

Figure 4.9. *Energy levels of Ce3 + in a YAG crystal: a) Mott diagram of Ce3 + excited at 460 nm, b) emission spectra of phosphor excited at 460 nm and c) Jablonski diagram of Ce^{3+} for two excitation wavelengths (340 and 460 nm). For a color version of the figure, see www.iste.co.uk/deshayes/reliability2.zip*

The Ce^{3+} atoms are replacing yttrium atoms in the $Y_3Al_5O_{12}$ host matrix. When the phosphor is not excited, the ΔR distance between the cerium (Ce) and oxygen (0) atoms is zero. The 4f and 5d levels of phosphor are aligned. When the phosphor is illuminated, the excitement of the Ce atom makes the distance ΔR nonzero ($\Delta R'$) and the energy bands 4f and 5d no longer aligned (Figure 4.9(a)).

The excitation spectrum of the YAG:Ce at 532 nm makes it possible to highlight the electronic transitions of levels $^4F_{5/2}$ towards the excited levels $^5D_0$ and $^5D_1$. Thus, the transition $^4F_{5/2} \rightarrow {}^5D_1$ gives an absorption peak centered at 340 nm and the transition $^4F_{5/2} \rightarrow {}^5D_0$ gives an absorption peak centered at 460 nm. The shape of the absorption peaks is Lorentzian according to the system theory of two discrete levels. By exciting the YAG:Ce at 340 nm and 460 nm, the phosphor's ^5D levels are excited and electronic transitions $^5D_0 \rightarrow {}^4F_{7/2}$ and $^5D_0 \rightarrow {}^4F_{5/2}$ lead to two emission bands that form the emission band from 2.1 to 2.4 eV (Figures 4.9(b) and (c)).

Figure 4.10 shows the spectrum of a CREE white LED technology using the YAG:Ce phosphor.

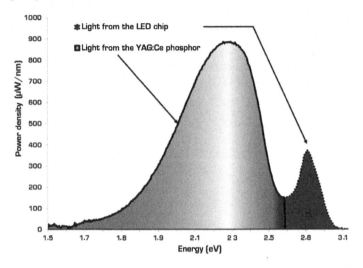

Figure 4.10. *Emission of a white LED CREE phosphor technology YAG:Ce. For a color version of the figure, see www.iste.co.uk/deshayes/reliability2.zip*

This emission spectrum is that of the silicone oil/phosphor mix excited by the light from the LED. We thus find the phosphor emission band of 2.1–2.4 eV and the LED emission band of 2.55–3.1 eV. The rest of the emission is due to the silicone oil fluorescence emission and extends to 1.7 eV.

The second effect is the loss of optical power of the light emitted by the LEDs. Figure 4.11 illustrates this phenomenon by comparing LEDs with (+P) and without phosphor before aging.

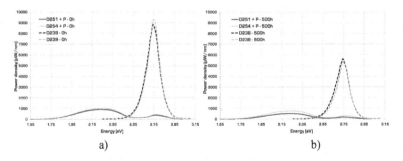

Figure 4.11. *Optical spectra of blue and white LEDs (+P): a) before aging and b) after aging*

Adding the phosphor thus indicates a significant loss of optical power (> 90%) to the central wavelength λ_C of the LEDs. Table 4.11 shows the impact of the addition of phosphor on the optical power of the LED before and after aging.

LED	$P_{optical\ chip}$ (W) $-$ 0 h	$P_{optical\ chip}$ (W) $-$ 500 h
D238	0.253	0.162
D239	0.253	0.144
D240	0.260	0.170
D241	0.232	0.185
D251 + P	0.017	0.008
D252 + P	0.018	0.009
D254 + P	0.019	0.014
Average blue optical losses (%)	92.80	94.14

Table 4.11. *Impact of the addition of the phosphor on the optical power of the chip LEDs before and after aging*

The values of the optical powers of the LEDs with phosphor correspond only to the chip's emission (from 2.6 to 3 eV). The average optical power loss at 500 h is increased by 1.34%. This means that, after aging, light absorption of LEDs in the blue (at λ_C) is particularly important.

The third effect of the addition of the phosphor is the spectral shift of the central wavelength of the LEDs. Figure 4.12 shows the normalized optical spectra on λ_C of LEDs before and after aging.

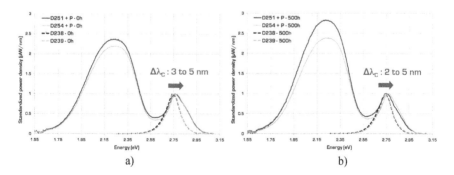

a)   b)

Figure 4.12. *Impact of the phosphor on the spectral shift of the λ_C parameter of LEDs: a) before aging and b) after aging*

The addition of phosphor involves a spectral shift of 3–5 nm of λ_C towards the blue before aging. The spectral shift could be due to an improvement in the thermal resistance (lower junction temperature), which is characterized by a shift of λ_C towards blue. Table 4.12 compares the values of the junction temperature, the thermal resistance and the slope of the linear portion of the T_J (I) curve of the two technologies before aging.

Technology	LED	T_J (°C) – 0 h	R_{th} (K.W^{-1}) – 0 h	Slope $_{400/700\ mA}$ T_J(I) – 0 h
Without phosphor	D238	59.49	40.35	0.292
	D239	60.88	40.48	0.321
	D240	67.07	49.54	0.317
	D241	57.58	36.45	0.324
With phosphor	D251	50.91	26.03	0.294
	D252	62.49	38.53	0.368
	D254	69.98	47.38	0.255

Table 4.12. *Junction temperature, thermal resistance and slope of the linear portion (400/700 mA) of the T_J (I) curve of the blue and white LEDs at 350 mA and T_P = 300 K*

These values show a significant dispersion of the thermal resistance R_{th} which does not allow concluding on the spectral shift. This seems to suggest a reproducibility problem of the assembly process (In$_{50}$Sn$_{50}$ weld).

At 500 h, the deviation of the spectral blue shift of λ_C widened from 2 to 5 nm and most λ_C of LEDs with and without phosphor were red shifted. Table 4.13 summarizes the central wavelengths λ_C of each chip (blue) before and after aging.

Conditions	Aging time (h)	$\lambda_{CLEDs\ without\ phosphor}$ (nm)				$\lambda_{CLEDs\ with\ phosphor}$ (nm)		
		D238	D239	D240	D241	D251	D252	D254
T_P = 300 K, I = 350 mA	0	452	451	451	451	448	448	447
	500	452	452	452	452	450	450	447
T_J = 300 K, I = 700 mA	0	445	443	444	444	441	440	442
	500	445	443	444	444	441	-	442

Table 4.13. *Shifts of central wavelength of LEDs at T_P and T_J = 300 K before and after aging*

After aging, the phenomenon is the same for the LEDs without phosphor: the central wavelengths at T_J = 300 K were unchanged while at T_P = 300 K, a difference of 2 nm towards red is present. Presumably, this spectral shift is due to a change in the heat flow within the assembly after aging as in the case of LEDs without phosphor. Figure 4.13 shows the current variation as a function of the junction temperature before and after aging at T_P = 300 K.

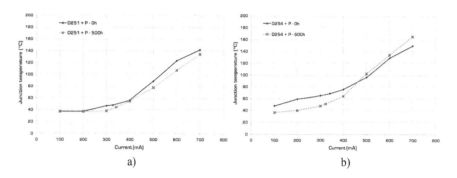

a)   b)

Figure 4.13. *TJ change as a function of current for white LEDs at 0 and 500 h: a) LED 251 and b) LED 254*

The T_J (I) characteristic shows an overall decrease (about 10°C) in the LEDs' junction temperature. This means that the thermal management of LEDs is improved after aging. Table 4.14 summarizes the varying thermal parameters extracted from the T_J (I) characteristic for a current of 350 mA at T_P = 300 K before and after aging.

LED	T_J (°C) – 0 h	T_J (°C) – 500 h	ΔT_J (°C)	ΔR_{th} (K.W^{-1})	gradual change of slope (%)
D251	50.91	45.90	−5.02	−7.00	−7.78
D254	69.98	55.75	−14.23	−17.47	23.96

Table 4.14. *Variations in the thermal parameters of the white LEDs at 350 mA and the slope of the linear portion (400 – 700 mA) of the T_J(I) curve at T_P = 300 K before and after aging*

As in the case of LEDs without phosphor, the decrease in the thermal resistance of LEDs confirms better thermal management. However, the central wavelength λ_C increased by 2 nm towards the red at T_P = 300 K after aging for most LEDs. This observation indicates that the temperature dependence of the center wavelength has been changed. To confirm this, we have estimated, from the literature, the slope of

λ_C as a function of T_J ($d\lambda_C/dT_J$) before aging. Figure 4.14 shows the change in λ_C as a function of T_J for the white LED 251 before and after aging.

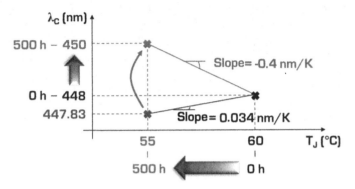

Figure 4.14. *Variation in λ_C as a function of T_J before and after aging for the white LED 251*

At 500 h, with an approximately 5°C loss of T_J and a slope of 0.034 nm.K^{-1} at the initial state, the λ_C central wavelength of the LEDs should be less than 448 nm. However, after aging, we saw it increased by 2 nm. The $d\lambda_C/dT_J$ slope has therefore become negative. Table 4.15 shows the impact of aging on the temperature dependence of the central wavelength of the LEDs without phosphor.

LED	$d\lambda_C/dT_J - 0$ h	$d\lambda_C/dT_J - 500$ h	Variation (%)
D251	0.034	− 0.399	91.47
D254	0.034	0.034	0.00

Table 4.15. *Impact of aging on the slope $d\lambda_C/dT_J$ for LEDs with phosphor*

These results demonstrate a thermal failure signature defined by a variation in the slope of the central wavelength of the LEDs in temperature after aging. This confirms the modification of the thermal flow of LEDs at 500 h, which results in a spectral shift of 2 nm and a reduction in junction temperature of 5 − 14°C at 350 mA and $T_P = 300$ K.

The component is thus thermally improved and should result in improved LED performance. Despite this, the optical power drops seen after aging are substantial. Figure 4.15 shows the optical spectra of white LEDs powered at 350 mA at $T_P = 300$ K before and after aging for each spectrum recalling the total optical power loss.

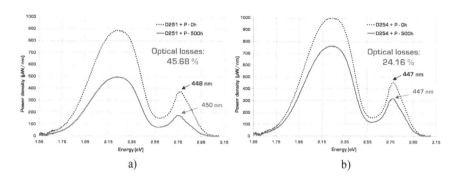

a)   b)

Figure 4.15. *Optical spectra L (E) of the white LEDs powered
at 350 mA at 0 and 500 h: a) LED 251 and b) LED 254*

At 500 h, the white LEDs have lost between 24 and 46% of their original optical power. The optical power loss is calculated for the entire spectrum and a digital integral difference representing the area bounded by the two curves L (E) at 0 h and L (E) at 500 h. In addition to the intrinsic losses due to the LED's luminescence, we must evaluate the yield loss of the frequency conversion of blue/yellow. To enable this study, one of the methods is to separate the analysis from the two luminescence peaks. The peak represented by λ_C will be associated with the absorption yield and the peak centered at 2.2 eV (yellow→$\lambda_{phosphor}$) will be associated with the phosphor's fluorescence yield. We will also evaluate the yellow/blue ratio ΔYBR. Table 4.16 summarizes all this information before and after aging.

Parameters	D251		D252		D254	
	0 h	500 h	0 h	500 h	0 h	500 h
YBR	6.896	8.394	6.902	8.061	6.850	7.295
ΔYBR (%)	0	21.72	0	16.79	0	6.50
Loss P$_{opt\ LED}$ (%)	0	54.49	0	48.26	0	28.62
Loss P$_{opt\ phosphor}$ (%)	0	44.60	0	39.58	0	23.98

Table 4.16. *Impact of aging on the yellow/blue
ratio and optical power loss in the blue and yellow*

The increase in the ΔYBR ratio and the comparison of the optical power losses in the two areas corresponding to λ_C (LED) and $\lambda_{phosphor}$ (phosphor fluorescence) peaks confirm that the LED's light optical absorption at the λ_C peak is greater than that of the $\lambda_{phosphor}$ peak. This absorption, due to the phosphor molecules of the mixture of silicone gel/phosphor is related to the white color drifting towards yellow. Optical power loss and yellowing of the white light can also be explained by the deterioration of the silicone oil's fluorescence yield, and also by the phosphor's aging. Indeed, Zhang *et al.* have studied the impact of temperature on the spectral characteristics of a YAG:Ce powder. The latter demonstrated that the light intensity (due to the optical power) is degraded by about 50% when the phosphor is subjected to a temperature of 453 K [ZHA 08]. However, this phenomenon must be reversible when the phosphor returns to room temperature. The aging temperature of 453 K is very close to the LED junction temperature during aging at 85°C (350 K).

Electrical, optical and thermal failure signatures of white LEDs lead to the following conclusions:

– Overall, the LEDs and their assembly (with or without phosphor) have, from a thermal point of view, improved at 500 h. Indeed, a decrease in thermal resistance is highlighted (3–17 K.W^{-1}), which resulted in a decrease in the junction temperature of the LEDs (0.5–14°C). Despite this, the observed optical power drops (20–46%) are considerable. In addition, the LEDs' central wavelength did not vary at $T_J = 300$ K, while, at $T_P = 300$ K, the latter shifted at 500 h by 1 and 2 nm towards red for the blue and white LEDs, respectively. The junction temperature having decreased after aging, the spectral shift at $T_P = 300$ K should have occurred towards the blue consistent with the variation in the gap of the active area in temperature. It has thus been highlighted that the LEDs' $d\lambda_C/dT_J$ slope became negative. This spectral shift made it possible to highlight a change in thermal flow of LEDs, which resulted in an improved device after aging. On the other hand, the spectral shift led us to hypothesize that the optical power loss is due to a phenomenon external to the chip. We thus suspected a degradation of the fluorescence of the silicone oil (with and without phosphor) induced by the thermal conditions during the LEDs' aging.

– Table 4.17 shows the LED's junction temperature before aging at 550 mA and $T_P = 300$ K and a T_J estimation during aging at $T_P = 358$ K (85°C).

The increase in the device's junction temperature during aging results, within the coating polymer, in a thermally activated degradation. Indeed, we can assume that the aging conditions create a molecular modification of the polymer and thus the phosphor's environment. This molecular change is induced by exceeding the glass transition temperature beyond which the polymer is liquid and can have its composition and structure modified. The often observed phenomena consist of the diffusion of the assembly materials surrounding the polymer/phosphor mix, precipitation of large molecules or the occurrence of occlusions due to temperature differences. Note also that, during this phase, the polymer is subjected to a power density of the LEDs' chip of about 1,480 $W.cm^{-2}$. The photo-thermal reaction demonstrated in Chapter 3 can also be activated in this type of device.

Conditions	LEDs without phosphor				LEDs with phosphor		
	D238	D239	D240	D241	D251	D252	D254
T_J (K) to T_P = 300 K	385.25	389.09	390.77	382.72	375.76	395.19	385.64
T_J (°C) to T_P = 300 K	112.10	115.94	117.62	109.57	102.61	122.04	112.49
T_J (K) to T_P = 358 K	443.25	447.09	448.77	440.72	433.76	453.19	443.64
T_J (°C) to T_P = 358 K	170.10	173.94	175.62	167.57	160.61	180.04	170.49

Table 4.17. *Junction temperature of the LEDs before (300 K) and during (358 K) aging*

To validate the failure mechanisms explaining the change in thermal flow of blue and white LEDs, the optical power loss and the shift of the color of the LEDs after aging, two types of physicochemical analyses will be used:

– A fluorescence analysis of the silicone oil alone (blue LEDs), then of the mixture (white LEDs) will confirm two points:

  - impact of aging on the fluorescence of the silicone oil using an excitation wavelength equivalent to that of the LEDs;

  - impact of aging on the modification of the silicone oil's molecular structure by exciting the latter in the UV;

– An analysis by X-ray diffraction on the silicone oil/phosphor mixture will help to identify whether the phosphor has been affected by aging.

4.3.2. Validation of failure mechanisms by using physiochemical analyses

We saw in the previous section that the LEDs lost more than 45% of their original optical power. The two most affected LEDs in each technology are: LED 239 without phosphor and LED 251 with phosphor. The study will mainly focus on the physicochemical characteristics of the silicone oils used for these two LED assemblies. For this, we established a strategy for comparing the oils used before and after aging despite the destructive nature of the physicochemical analyses. The assembly polymers were therefore prepared using a part for the assembly of LEDs and the other part as a reference for physicochemical analyses. Therefore, the comparisons made in this chapter are, for the first time in terms of reliability study, perfectly absolute and referenced.

We will use the various analyses presented in Chapter 2 to determine the difference in chemical composition, environment of the phosphor and electronic transition properties (absorption and emission). The association with Bordeaux chemistry laboratories ISM (organic materials) and ICMCB (inorganic materials) for this research helped give a scientific validity with experts from each of the specialized areas.

4.3.2.1. Fluorescence analysis of silicone oil without phosphor

The fluorescence analysis was performed on the chip's silicone oil coating with an excitation wavelength of 450 nm (2.76 eV) corresponding to the central wavelength of the LEDs. This is the starting point of the fluorescence analysis. Figure 4.16 shows the normalized fluorescence emission spectrum before aging when the oil is excited at 450 nm and the corresponding Jablonski diagram.

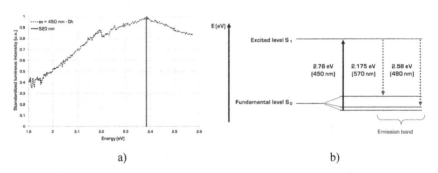

a)   b)

Figure 4.16. *a) Fluorescence emission spectrum at $\lambda_{excitation}$ = 450 nm and b) the corresponding energy diagram (PDMS)*

We denote the fundamental level of a PDMS molecule by the level singlet S_0 and the excited level without spin rotation by singlet level S_1. By exciting the silicone oil at the central wavelength of the LEDs, we confirm an absorption ($^0S_0 \rightarrow S_1$) of the light emitted by the chip at 450 nm and a fluorescence emission of the oil of 480 nm ($S_1 \rightarrow {}^2S_0$) at 570 nm ($S_1 \rightarrow {}^1S_0$) with a maximum peak at 520 nm (2.38 eV). Figure 4.17 shows the silicone oil's excitation spectrum at 520 nm.

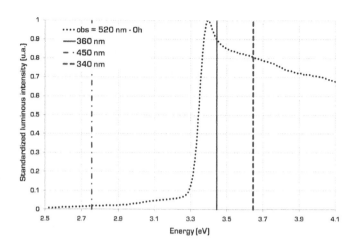

Figure 4.17. *Excitation spectrum at λ_observation = 520 nm*

This excitation spectrum shows a strong absorption of phosphors in the UV range at 360 nm (3.44 eV) and 340 nm (3.65 eV). There is also a phosphor absorption peak at 450 nm (2.76 eV). Figure 4.18 shows the fluorescence emission spectra achieved by exciting the PDMS at 360 and 340 nm and the corresponding Jablonski diagrams.

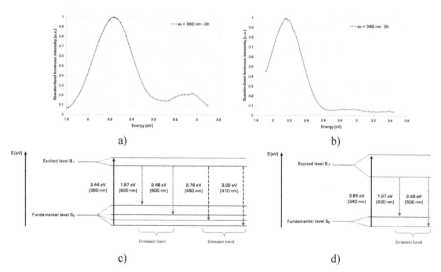

Figure 4.18. *a) Emission spectrum at λ_excitation = 360 nm, b) emission spectrum at λ_excitation = 340 nm, c) Jablonski diagram at λ_excitation = 360 nm and d) Jablonski diagram at λ_excitation = 340 nm*

The major advantage of using a UV excitation source (360 nm) is to observe a fluorescence spectrum revealing possible modifications in the PDMS molecular structure that may be related to the appearance of small or large molecules. We find a fluorescence emission from 500 to 630 nm for absorption (only phosphors) at 340 and 360 nm. The complementarity of these two spectra is the information given by the emission spectrum when the oil absorbs at 360 nm: a second fluorescence emission band appears from 410 to 450 nm.

Table 4.18 gives the fluorescence yield of the silicone oil's molecules as a function of the wavelength ranges before aging.

$\lambda_{absorption}$ (nm)	$I_{absorption}$ (u.a.) – 0 h	$\lambda_{fluorescence}$ (nm)	$I_{fluorescence}$ (u.a.) – 0 h	$\eta_{fluorescence}$ (%) – 0 h
340	$1.78.10^7$	575	$4.23.10^6$	23.80
360	$1.98.10^7$	555	$6.43.10^6$	32.37
		418	$1.39.10^6$	6.99
		431	$1.33.10^6$	6.68
450	$4.20.10^5$	520	$3.91.10^5$	93.07

Table 4.18. *Silicone oil's fluorescence yields before aging*

These values indicate a very good fluorescence yield (93%) for absorption centered at 450 nm although the light intensity is lower by two decades in absorption. This indicates that the frequency conversion between 450 nm and the 520–630 nm bands is very efficient and provides a wide yellow peak (Figure 4.10). However, the absorption efficiency is controlled so that the blue–yellow mixture corresponds to a white CRI that complies with industrial requirements. This analysis allows establishing the link between the CRI performance and the material's composition. This chemical composition is available by fluorescence analyses in the UV band of 340–360 nm. These are smaller but much more intense in fluorescence emission (2–5 decades).

Figure 4.19 shows the emission spectrum at 450 nm before and after aging.

Aging in active storage resulted in a loss of fluorescence by a factor of 10 (90%). In addition, a spectral shift of about 20 nm towards the red, from 500 to 590 nm (2.48–2.10 eV), was observed at 500 h. This is consistent with the spectral shift of the LEDs towards the red (from 451 to 452 nm). The maximum fluorescence emission is now at 540 nm. Figure 4.20 shows the silicone oil's excitation spectrum at $\lambda_{observation}$ = 520 nm prior to aging and $\lambda_{observation}$ = 540 nm after aging.

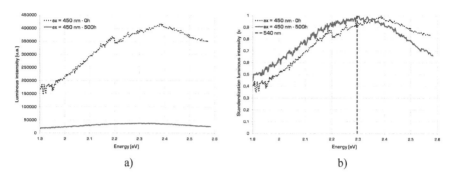

Figure 4.19. *a) Fluorescence emission spectrum at $\lambda_{excitation}$ = 450 nm at 0 and 500 h and b) fluorescence emission spectrum at normalized $\lambda_{excitation}$= 450 nm at 0 and 500 h*

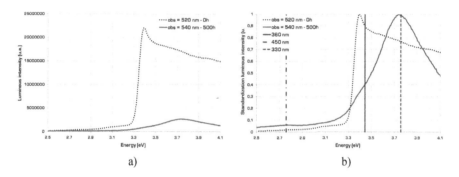

Figure 4.20. *a) Excitation spectra at $\lambda_{observation}$ = 520 nm (0 h) and $\lambda_{observation}$ = 540 nm (500 h) and b) standardized excitation spectra at $\lambda_{observation}$ = 520 nm (0 h) and $\lambda_{observation}$ = 540 nm (500 h)*

After aging, absorption towards UV wavelengths (360 nm and 330 nm) became predominant, although, at 450 nm, it still exists but becomes very low. Table 4.19 gives the absorption losses for the three excitation wavelengths of 330, 360 and 450 nm at 500 h.

Aging time (h)	Losses at 330 nm (%)	Losses at 360 nm (%)	Losses at 450 nm (%)
500	84,99	94,62	62,76

Table 4.19. *Losses of absorption of phosphors after aging at 340, 360 and 450 nm*

At 450 nm, the optical absorption of phosphors present in the silicone coating decreased by over 60%, which explains the significant loss of fluorescence. To explain the origin of such an absorption loss, fluorescence emission spectra were performed by exciting in the UV (360 nm). This allows studying the changes in the molecular structure of PDMS. Figure 4.21 shows the fluorescence emission spectra at 360 nm before and after aging.

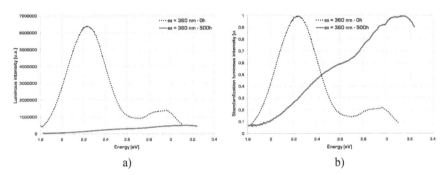

a)   b)

Figure 4.21. *a) Fluorescence emission spectrum at $\lambda_{excitation}$ = 360 nm at 0 and 500 h and b) standardized fluorescence emission spectrum at $\lambda_{excitation}$ = 360 nm at 0 and 500 h*

The fluorescence emission spectrum, by exciting the PDMS at 360 nm, indicates an almost inverted fluorescence. Indeed, when standardizing this spectrum (Figure 4.21(b)), the highest fluorescence peak is located in the areas of lower wavelengths 410 – 380 nm (3.02 – 3.26 eV) with other peaks at 420 nm (2.95 eV) and 450 nm (2.76 eV). Before aging, there was a large fluorescence peak between 500 and 630 nm (2.48 and 1.97 eV). We have observed for a similar oil in Chapter 3, the increase in fluorescence in the low energy that is related to a change of the silicone oil's molecular structure. In addition, the RMN ^{1}H dosimetry analysis confirmed the existence of LM. These two elements could be related. If we refer to the fluorescence analysis performed in Section III and the literature, the increase in fluorescence in the high energies indicates that there is a change in the silicone gel's molecular structure, eventually connected to an increase in the concentration of SM after aging. In the case of this chapter, the phenomenon of the modification of the molecular structure of the gel is also confirmed by a second UV fluorescence analysis, at the excitation wavelength of 330 nm. Figure 4.22 shows the fluorescence emission spectra by exciting the PDMS at 330 nm before and after aging.

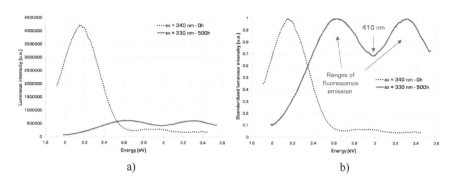

a)   b)

Figure 4.22. *a) Fluorescence emission spectrum at* $\lambda_{excitation}$ *= 330 nm at 0 and 500 h and b) standardized fluorescence emission spectrum at* $\lambda_{excitation}$ *= 330 nm at 0 and 500 h*

Two fluorescence emission ranges appear very clearly in the shorter wavelengths (high-energy): from 350 to 410 nm (from 3.54 to 3.02 eV) and from 410 to 430 nm (from 3, 02 to 2.88 eV). This mechanism confirms our interest in exciting the silicone oil with UV radiation. A Si oil modification mechanism appears after aging and shifts the fluorescence spectrum towards lower wavelengths. This molecular change results in a shift of fluorescence towards red by exciting at the LED's emission wavelength, or at 450 nm. It may be reasonable to assume that this change could mean breaks in molecular chains leading to an increase in the concentration of SM.

Table 4.20 gives the fluorescence yield of the silicone oil's molecules excited at 450 nm as a function of wavelength ranges.

$\lambda_{absorption}$ (nm)	$I_{absorption}$ (u.a.) – 500 h	$\lambda_{fluorescence}$ (nm)	$I_{fluorescence}$ (u.a.) – 500 h	$\eta_{fluorescence}$ (%) – 500 h
450	$1.56.10^{5}$	540	$3.72.10^{4}$	23.81

Table 4.20. *Fluorescence yield of the silicone oil excited at 450 nm after aging*

At 450 nm, the fluorescence yield is decreased from 93% at 0 h to 24% at 500 h. This 69% decrease is one reason for LEDs' optical power loss.

The optical power loss of the blue LEDs is thus due to a change in the molecular structure of the silicone oil activated by a photo-thermal mechanism. The modification of the polymer structure may result in breaks in molecular chains leading to the appearance of SM (increase in their concentration). This failure mechanism results in a modification of the phosphor environment and therefore of the electronic transitions. We then observe a spectral red-shift (20 nm in fluorescence and 1 nm at the LED's output), and a decrease of 69% of the fluorescence yield at 450 nm due to phosphors' reduced absorption of 63%. Figure 4.23 shows a top view, taken using an optical microscope, of the coating polymer of blue LEDs before and after aging.

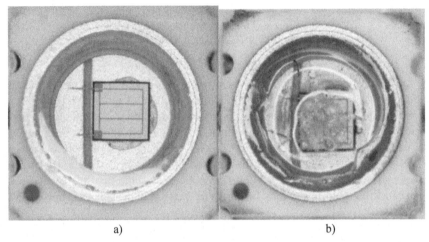

a)   b)

Figure 4.23. *Optical microscope image of the top view of the coating polymer of a blue LED: a) before aging and b) after aging*

The visual appearance of the polymer after aging confirms that there has been a change in the oil's molecular structure, creating cracks or bubbles in some places (area above the chip). We can hypothesize an assembly method problem or incompatibility with the finish of the package or transfer material to the silver glue.

4.3.2.2. Fluorescence and X-ray diffraction analysis of the silicone oil with phosphor

The addition of the phosphor in the silicone oil plays a major role in the fluorescence emission of the latter. Figure 4.24 shows the standardized fluorescence spectrum at an excitation wavelength of 450 nm before aging as well as the corresponding Jablonski diagram.

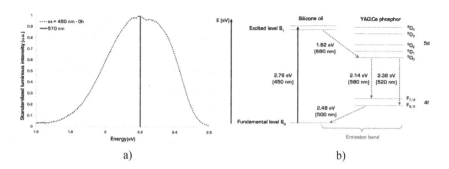

a)   b)

Figure 4.24. a) Fluorescence emission spectrum at $\lambda_{excitation}$ = 450 nm at 0 h and b) the corresponding Jablonski diagram (PDMS + YAG:Ce)

The emission spectrum at 450 nm is very uniform. It is a large spectrum centered at 570 nm (2.175 eV). It therefore confirms the presence of fluorescence of the PDMS/YAG:Ce mixture (dosage 85%/15%) when the LED is powered. When observing at 570 nm, we find a section of the YAG:Ce absorption spectrum observed by Tomiki et al. [TOM 91]. Figure 4.25 shows this excitation spectrum before aging.

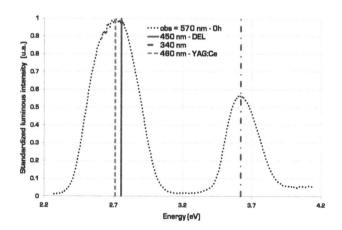

Figure 4.25. Excitation spectrum of the PDMS/ YAG:Ce mixture at $\lambda_{observation}$ = 570 nm before aging

We find the two absorption peaks at 460 and 340 nm corresponding to electronic transitions $^4F_{5/2} \rightarrow ^5D_1$ and $^4F_{5/2} \rightarrow ^5D_0$ of the YAG:Ce, respectively. The LED's wavelength, 450 nm, is therefore very close to the absorption maximum at 460 nm. This confirms the phenomenon of absorption of light emitted by the chip through the silicone oil/phosphor mixture. Figure 4.26 shows the emission spectrum of the coating with phosphor, and its corresponding energy diagram, before aging and excited at 340 nm.

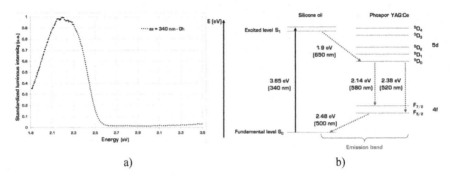

a)   b)

Figure 4.26. *a) Fluorescence emission spectrum at λ_{excitation} = 340 nm at 0 h and b) the corresponding Jablonski diagram (PDMS + YAG:Ce)*

The spectral range extends here from 500 to 650 nm and remains greater than twice the spectral range of the YAG:Ce.

After aging, the fluorescence of the silicone oil has fallen sharply. Figure 4.27 shows the fluorescence emission spectra (actual and standardized at 1) of the coating phosphor excited at 450 nm before and after aging.

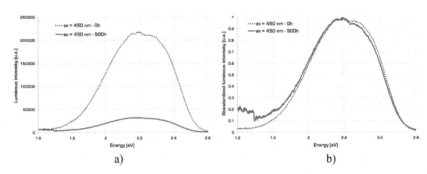

a)   b)

Figure 4.27. *a) Fluorescence emission spectrum at λ_{excitation} = 450 at 0 and 500 h and b) standardized fluorescence emission spectrum at λ_{excitation} = 450 at 0 and 500 h*

The spectral range remained the same but the fluorescence intensity decreased by 85%. This loss is mainly due to the PDMS. Indeed, at a junction temperature of around 450 K, Zhang *et al.* reported a fluorescence loss of 50%. However, without phosphor, it was shown that, at 450 nm, the fluorescence decreases by 90%

after the same aging. With a dosage of 85% of PDMS and 15% of phosphor, when adding up the part of the losses of each component, 85% of 90% of losses for the PDMS and 15% of 50% losses for the YAG : Ce, we get 85% fluorescence loss. This calculation is true in aging since the 50% loss of fluorescence of the YAG:Ce is a normally reversible phenomenon.

Figure 4.28 shows the excitation spectrum (actual and standardized at 1) of PDMS/YAG:Ce mixture before and after aging.

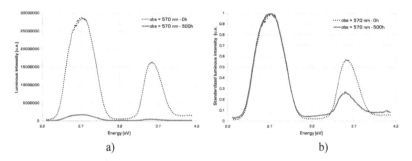

a)   b)

Figure 4.28. *a) Excitation spectrum of the mixture of PDMS/ YAG:Ce at $\lambda_{observation}$ = 570 nm at 0 and 500 h and b) standardized excitation spectrum of the mixture of PDMS/ YAG: Ce at $\lambda_{observation}$ = 570 nm at 0 and 500 h*

The main impact of aging on the performance of the phosphor coating corresponds to a loss of absorption of the silicone oil's phosphors of 94% at 450 nm. To understand the physical origin of this absorption loss, the fluorescence spectrum of the PDMS/YAG:Ce mixture was performed by excitation with UV radiation. Figure 4.29 shows the phosphor coating's emission spectrum excited at 345 nm before and after aging.

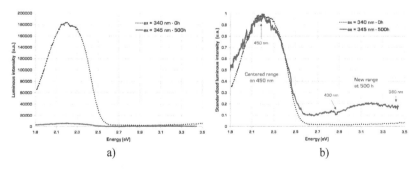

a)   b)

Figure 4.29. *Fluorescence emission spectra: a) $\lambda_{excitation}$ = 345 nm at 0 and 500 h and b) $\lambda_{excitation}$ = 345 nm standardized at 0 and 500 h*

The same mechanism of fluorescence increase at the highest energies is present. Indeed, a field emission is observed from 360 to 430 nm (3.44 – 2.88 eV) of the silicone oil in addition to the centered range at 450 nm (2.48 – 1.9 eV) interacting with the phosphor. The modification mechanism of the silicone oil's molecular structure is present for the white LEDs at 500 h. The latter could be related to the molecular chain breaks that could lead to an increase in the concentration of SM. The addition of the phosphor reduces this mechanism, but does not prevent fluorescence absorption/re-emission mechanisms in silicone oil. Table 4.21 summarizes all losses of absorption and fluorescence emission, and compares fluorescence yields before and after aging.

Aging time (h)	$\eta_{fluorescence}$ (%)		Absorption losses (%)		Fluorescence losses (%)	
	340 nm	450 nm	340 nm	450 nm	340 nm	450 nm
0	1.16	0.77	0	0	0	0
500	1.37	1.92	97.37	94.06	96.89	85.23

Table 4.21. *Fluorescence yield, loss of absorption and fluorescence emission of the LEDs with phosphor coating before and after aging*

At 500 h and at the LEDs' wavelength (450 nm), the polymer/phosphor mixture is more efficient. Indeed, at 450 nm, the fluorescence yield rose 1.15% despite significant losses in both phosphor absorption (> 94%) and re-emitting fluorescence (> 85%). These losses are due to a modification mechanism of the molecular structure of the silicone oil leading to the loss of optical power of the LEDs and shift of their color towards yellow. The latter is due to a spectral shift in the UV fluorescence of 5 nm towards the blue that causes a shift of 2 nm towards the red in the light at the encapsulated LED's output. The modification of the Si oil's molecular structure can be linked to a mechanism of breaking the molecular chains that may lead to the increase in the concentration of SM.

Figure 4.30 shows a view from above, taken by optical microscopy, of the coating polymer of a white LED before and after aging.

The visual appearance of the polymer after aging confirms that there has been a change in the molecular structure of the oil, creating even cracks in some areas. We can also make the same hypotheses for the blue LEDs, related to an assembly method problem or incompatibility with the finishing of the package or transfer material to the silver glue.

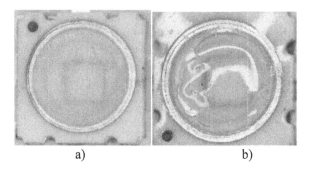

a)   b)

Figure 4.30. *Optical microscopic image of the top view of the coating polymer in a white LED: a) before aging and b) after aging*

To verify the impact of aging on the YAG:Ce phosphor, we conducted an analysis by X-ray diffraction of several samples:

– a un-aged sample (silicone oil/phosphor mix);

– a sample (silicone oil/phosphor mix) aged for 500 h;

– a sample without phosphor aged for 500 h.

Figure 4.31 presents the diffractograms of these three samples with a diagram of the sample structure.

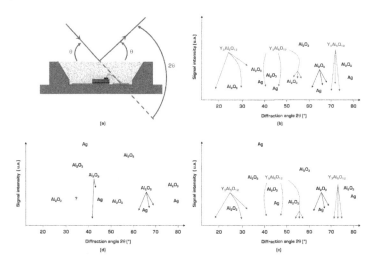

Figure 4.31. *a) Diagram of the structure of the analyzed samples, b) coating with YAG:Ce - 0 h, c) coating with YAG:Ce - 500 h and d) coating without YAG:Ce - 500 h. For a color version of the figure, see www.iste.co.uk/deshayes/reliability2.zip*

The sample without phosphor aged at 500 h makes it possible to distinguish the rays coming of the assembly (Ag and Al_2O_3) from those of the YAG:Ce ($Y_3Al_5O_{12}$). After aging, a reduction (<90%) was observed of $Y_3Al_5O_{12}$ rays without any appearance of other rays. The decomposition of the YAG:Ce phosphor can be exposed in the forms given by equations [4.1] and [4.2].

$$Y_3Al_5O_{12} \rightarrow \frac{3}{2}Y_2O_3 + \frac{5}{2}Al_2O_3 \qquad [4.1]$$

$$Y_3Al_5O_{12} \rightarrow 3YAlO_3 + Al_2O_3 \qquad [4.2]$$

In our case, the rays of Al_2O_3 do not suggest a degradation of YAG:Ce as the analyzed sample is mainly composed of the ceramic substrate (Al_2O_3). Furthermore, no additional ray containing yttrium (Y_2O_3 or $YAlO_3$) appears after aging. This means that the YAG:Ce phosphor has not decomposed. The hypothesis to explain the high attenuation of the $Y_3Al_5O_{12}$ rays is the chemical interaction between the silicone oil and the phosphor. Indeed, a surface diffusion of yttrium atoms in the silicone oil could be related to the high attenuation of $Y_3Al_5O_{12}$ rays.

4.3.3. Technological solutions

To solve the silicone oil degradation problems related to aging in applied active storage, several solutions can be implemented using the same chip:

– With the YAG:Ce phosphor, the color rendering index is often low ($\approx$ 75) because of the sharp contrast between the luminous intensity in the red field and the blue. To improve the CRI and light extraction, Won et al. proposed the use, from a blue LED, of two phosphors separated by a layer of silicone oil: (Ba, Sr) $2SiO_4$: Eu^{2+} green and $CaAlSiN_3$: Eu^{2+} red. This has led to a CRI of 95 and a luminous efficiency of 51 lm.W^{-1} at 350 mA;

– Different geometries of phosphor and silicone oil have been tested over the past decade. When the phosphor is in direct contact with the chip, an optical power loss by absorption of 60% is effective. When the latter is separated and positioned above using smooth reflective layers, the luminous efficiency is improved by 36% [KIM 05b]. By using diffuse reflective layers, the efficiency is improved by 75%. Furthermore, if we add a spherical lens above the phosphor layer, the yield gain is again 20% [LUO 05];

– Creating copper thermal vias of 0.8 mm diameter in the FR4 layer can reduce the thermal resistance of the assembly to 4°C.W^{-1} [CRE 10].

Finally, by changing the chip and replacing the silicon substrate with a copper substrate, Lau *et al.* have reported an increase in the optical power of 80%.

4.4. Summary of results and conclusions

In this chapter, we highlighted the critical areas and degradation mechanisms intervening in aging in active storage (85°C/550 mA/500 h) for blue and white power LEDs via a methodology failure analysis. The main objective of this chapter was to show that the methodology developed on marketed components (Chapter 3) is integrated at the component's design phase to determine the most adapted technological choices to the environment for which the component is built. This is why this study was conducted on many components: with or without phosphor in the coating polymer.

Therefore, it has been shown in a first location phase of degraded areas that the optical power loss observed after aging (more than 45%) is due to the assembly, and more specifically, to the silicone coating with or without phosphor of power LEDs. Using electrical, thermal and optical failure signatures, we have shown a modification of the heat flow of the LEDs, which has the consequence of improving the device after aging. This resulted in a decrease in the thermal resistance (3–17 $K.W^{-1}$), which in turn resulted in a decrease in the LEDs' junction temperature (0.5–14°C). In addition, the LEDs' central wavelength shifted at 500 h, and $T_P = 300$ K, 1 and 2 nm towards the red for the blue and white LEDs, respectively. It was highlighted that the LEDs' $d\lambda C/dT_J$ slope turned negative. It was therefore shown that the loss of LEDs' optical power is due to a phenomenon external to the chip. Indeed, considering the study in Chapter 3, it was hypothesized that the junction temperature measured at 175°C during aging, and the light emitted by the chip (power density $\approx$ 1,480 $W.cm^{-2}$) played a role in the silicone oil fluorescence emission.

To confirm the degradation of the silicone oil with and without phosphor, fluorescence analysis was performed for each type of LED. This analysis demonstrated that the silicone oil is degraded by a modification mechanism of the Si oil structure enabled by two aggravating factors: the junction temperature during aging and the chip's light at 450 nm. This modification resulted in:

– A spectral red-shift (20 nm in fluorescence and 1 nm at the LED's output) for blue LEDs. A spectral shift of 5 nm of the UV fluorescence towards the blue that causes a shift of 2 nm towards the red in the light at the white LEDs' output. These spectral shifts were used to explain the drift from 1 to 3.6% of white light towards yellow;

– A decrease of 69% in the fluorescence yield at 450 nm due to reduced absorption of phosphors (SM) of 63% for blue LEDs. Loss (> 85%) of fluorescence emission and absorption of SM of the silicone oil (> 94%) of the light emitted by the chip at 450 nm for white LEDs. These losses explain the optical power losses observed at the LEDs' output.

An assembly process problem or incompatibility with the finishing on the packaging or transfer material at the silver glue was highlighted and has confirmed the modification of the molecular structure of the coating polymer in view of the pictures taken under an optical microscope.

Degradation of the phosphor was studied by X-ray diffraction analysis. The results indicated that the YAG:Ce phosphor was not decomposed after aging. However, attenuation (> 90%) of the $Y_3Al_5O_{12}$ rays was observed. A surface diffusion of atoms of yttrium in the silicone oil could be related to the high attenuation of the $Y_3Al_5O_{12}$ rays.

Conclusion

This book presents the basics of the failure analysis methodology for assessing the reliability of GaN Leds.

Foremost, the interest in realizing electro-optical and thermal nondestructive characterizations was justified. The characterization benches of current–voltage, optical power, optical spectrum and the junction temperature measurement of an emitting optoelectronic system allow discriminating degradations between those of the chip and its assembly and pre-locating degraded areas. The second section has highlighted the interest of physicochemical analyses in identifying degraded areas of the assembly and determining the causes of failure. This book consists of four chapters.

In the first chapter, we presented state-of-the-art GaN technologies by focusing on three main points: the market for LEDs and GaN technologies, nitride materials and their binary and ternary cousins, and low and high power GaN components with associated assemblies. It has been shown that the evolution of these technologies led to increasingly complex and miniature developments for an increasingly common use.

The synthesis of all GaN technologies has allowed situating our study within a position that is international, national and within the framework and concepts developed by the EDMINA research team:

– the qualification standards are frequently used by all manufacturers in the field (OSRAM, Nichia, Cree, Philips Lumileds, Ledman, etc.);

– statistical and mathematical methods predict the operational lifetime of the component integrated with the system;

– the approaches of failure signatures, physical simulations and statistics were developed within the EDMINA team;

– a new concept of reliability integration was established by the EDMINA team to meet the fundamental industrial challenges.

The chosen approach is thus that of a methodology based on failure analysis and extraction of electrical and optical signatures to locate the failure. The addition of physicochemical analyses provides the opportunity to confirm the degradation mechanisms induced by aging in active storage.

In the second chapter, all of the tools and methods necessary for the encapsulated LEDs failure analysis were discussed. Four major tools were identified:

– the thermal characterizations allow extracting, via two methodologies, thermal parameters associated with the studied components from which the thermal models were built. An electrical method is preferably used for high-power LEDs in order to build the equivalent thermal models from the electrical/thermal analogy. The second proposed method is an optical method allowing the assessment of the junction temperature. This method has allowed the realization of the spectral characteristics at $T_J = 300$ K for low-power LEDs;

– the current–voltage I(V) characteristics were obtained using the KEITHLEY 6430 femto-ammeter. The I(V) characteristics gave access to the electrical parameters of each studied structure. This helped to build the equivalent electrical models from elementary electric dipoles: resistance and PN junction diodes. From electric models, an extraction of electrical failure signatures could be realized for each type of LED, allowing us to locate the degraded areas. All electrical measurements were carried out at constant temperature regulated by a liquid nitrogen BT500 cryostat controlled by a temperature regulator;

– the optical characteristics: the optical power using an OPHIR NOVA II measuring the latter from a UV photodiode, and the spectral characteristic using the TRIAX320 monochromator. The temperature of these measurements has been controlled by using the same thermal regulation bench used for electrical measurements. These characteristics have thus allowed determining the optical parameters associated with each structure, and the main operational parameter of an LED defined by the optical power. An optical model was implemented by taking the optical parameters related to the active area material;

– the physicochemical analyses in collaboration with many national laboratories. Two main issues were identified in this section: the significant contribution of information on the materials constituting the chip and its assembly, and the contribution of analyses that are specific to polymers in order to characterize the chip's coating material from the point of its optical function. Thus, a reminder of the principle of each analysis was proposed by emphasizing the importance of each one of these depending on the areas and/or materials to be analyzed.

The first study, developed in Chapter 3, highlighted the sensitive areas of low-power LEDs when subjected to aging in operational conditions (active storage: 1,500 h/85°C/I_{rated}). MQW InGaN/GaN blue emission LEDs (472 nm) were analyzed.

After aging, it was seen that the GaN LEDs underwent a loss of optical power of 65% for all LEDs at 1,500 h. Electrical failure signatures have failed to explain the drop in optical power measured at the output of the assembled component. However, a shift of the central wavelength of 3 nm towards blue was observed by optical spectra analysis at 1,500 h. Moreover, the central wavelength became insensitive to temperature. This phenomenon has allowed pre-locating the failure in the silicone coating oil by suspecting a shift of the fluorescence of the latter after aging. To explain the phenomena localized by electrical and optical failure signatures, we conducted several physicochemical analyses. The fluorescence analysis of the silicone oil excited in the UV (360 nm) has demonstrated two main phenomena:

– a luminous intensity inversion of the two maximum peaks (415 and 435 nm before aging) with spectral shift of about 3 nm for both peaks. The predominant shift being one showing shift towards UV (415 nm → 412 nm) in accordance with the same spectral shift (464 nm → 461 nm) observed on the LED's optical spectrum after aging;

– a 60% increase in the fluorescence luminous intensity with a maximum at 564 nm. In this case, we have assumed the presence of high-molecular-mass molecules (large molecules: LM).

An absorption loss higher than 90% at 1,500 h was also determined when the silicone oil is illuminated by the LED's light (464 m). This absorption is only the molecules re-emitting fluorescence (phosphors). Indeed, the optical silicone oil's absorption comprises two types of absorptions: an absorption of phosphor molecules and absorption of non-phosphor molecules. In this study, only the absorption of phosphor molecules was observed by means of fluorescence excitation spectra. The origin of the deterioration of the silicone oil's fluorescence was shown by three physicochemical analyses (DSC, RMN[1]H and MALDI-TOF mass spectrometry) confirming a modification mechanism of the silicone oil's molecular structure activated by the photo-thermal effect.

The results of the DSC were used to validate the process of the modification of the molecular structure of the Si oil indicating a temperature peak at 217°C. MALDI-TOF mass spectra and RMN spectra showed the disappearance of small molecules (old phosphors) and the existence of LM (new phosphors) after aging.

The Si oil's molecular structure modification could therefore be linked to a polymerization or cross-linking mechanism leading to an increase in the concentration of LM. Finally, the discrimination of the effect of temperature and light was conducted by a complementary fluorescence analysis on several silicone oil samples aged with or without light (464 nm). This has helped highlighting that light is partly responsible for the increase in the fluorescence emission in the red (564 nm), which can be related to the appearance of LM, while the temperature plays a role in the spectral shifts at high energies.

In the last chapter of this book, we highlighted the critical areas and degradation mechanisms intervening in aging in active storage (85°C/550 mA/500 h) of blue and white power LEDs through a failure analysis methodology. This chapter's main objective was to show that the methodology developed for marketed components (Chapter 3) is integrated at the design phase of a component to help manufacturers determine the most suitable technological choices for the environment for which the component is constructed. That is the reason why we conducted this study in component sets: with or without phosphor in the coating polymer.

Therefore, it has been shown in a first location phase of degraded areas that the optical power loss observed after aging (more than 45%) is due to the assembly, and more specifically, to silicone coating with or without phosphor for power LEDs. Using electrical, thermal and optical failure signatures, we have shown a modification of the heat flow of the LEDs, which has the consequence of improving the device after aging. This resulted in a decrease in the thermal resistance (3–17 K.W^{-1}), which resulted in a decrease in the junction temperature of the LEDs (0.5–14°C). In addition, the LEDs' central wavelength shifted at 500 h, and $T_P = 300$ K, by 1 and 2 nm towards red for blue and white LEDs. It has been highlighted that the $d\lambda_C/dT_J$ slope for LEDs became negative, showing that the LEDs' loss of optical power is linked to a phenomenon that is external to the chip. This first observation is justified by the conditions of aging related to environmental constraints, leading to a junction temperature of 175°C and a power density of 1,480 W.cm^{-2}.

To confirm the degradation of the fluorescence of the silicone oil with and without phosphor, a fluorescence analysis was performed for each type of LED. This analysis demonstrated that the silicone oil is degraded by a mechanism of molecular structure modification of the Si gel enabled by two contributing factors: the chip's junction temperature during aging and the chip's light at 450 nm. This modification resulted in:

– a spectral shift towards red (20 nm in fluorescence and 1 nm at the LED's output) for blue LEDs. A spectral shift of 5 nm of the UV fluorescence towards blue causes a shift of 2 nm towards red at the white LEDs' output. These spectral shifts were used to explain the shift of 1–3.6% of white light towards yellow;

– a decrease of 69% of the fluorescence yield at 450 nm due to reduced absorption of phosphors (SM) of 63% for blue LEDs. There are losses (>85%) of fluorescence emission and absorption of SM of silicone oil (>94%) for the light emitted by the chip at 450 nm for white LEDs. These losses explain the optical power losses observed at the LEDs' output.

The Si oil's molecular structure modification can be linked to a breaking mechanism of the molecular chains that may lead to the increase in SM concentration.

The phosphor's degradation was studied by X-ray diffraction analysis. The results indicated that the YAG:Ce phosphor was not decomposed after aging. However, attenuation (>90%) of the $Y_3Al_5O_{12}$ rays was observed. A surface diffusion phenomenon of the yttrium atoms in the silicone oil has been proposed as a hypothesis to justify the high attenuation of $Y_3Al_5O_{12}$ rays.

Images, taken under an optical microscope, of the polymer assembly of blue and white LEDs confirmed the modification of the molecular structure of the silicone oil. Cracks or bubbles were observed after aging in the Si oil. It was hypothesized that these could be related to a problem in the method of assembly or incompatibility with the finishing of the casing or transfer material to the silver glue.

Several technological solutions have been proposed to the manufacturer at the end of this analysis. At the assembly's level, the use of other phosphors and different geometries between the phosphor and the silicone oil were the main proposed solutions. A modification of the chip by replacing the silicon substrate by a copper substrate was also suggested.

We thus showed, through the methodology proposed in this book, how to conduct a failure analysis on optoelectronic components using electrical and optical failure signatures and, when necessary, physicochemical analyses.

The starting point of this methodology is the definition of environmental constraints from the mission profile. The latter allows, in particular, establishing the specifications of the aging tests. Aging analyses allow extracting electrical and optical failure signatures. These serve to identify degraded areas of the studied component and guide the physical and chemical analyses. The integration of physicochemical analyses represents the methodology's added value because it confirms degraded areas and explains the physical failure mechanism(s) involved in the degradation process.

All these steps will allow, in future research, to establish physical laws of degradation based on the degradation kinetics. The diagram of the proposed failure analysis methodology is shown in the figure below.

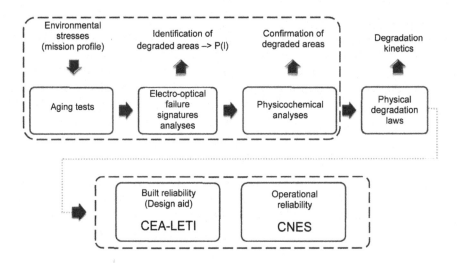

Figure 1. *Summary of the proposed methodology for failure analysis of encapsulated LEDs*

Further analysis by X-ray diffraction is one of the perspectives of this work by studying samples of YAG:Ce phosphors with and without silicone gel, extracted from the assembly of white LEDs and subjected to the same aging conditions in active storage.

On the other hand, a study of the silicone oil's total absorption, for both low and high-power LEDs, allows quantifying the share of LED light absorbed by the silicone oil, which does not retransmit in fluorescence. This would confirm the role of absorbing molecules other than phosphors.

It would be interesting to conduct a Raman spectroscopy analysis, completed by an IR analysis, on the silicone oil for low-power LEDs (Chapter 3) in order to determine the physical mechanism related to the modification of the oil's molecular structure: polymerization or cross-linking? Furthermore, a steric exclusion chromatography would, from the mass changes, confirm not only the latter, but also the chain-breaking process assumed in the study of the silicone gel's degradation conducted in Chapter 4 of this book.

In the longer term, our approach is part of a broader framework by adapting to other types of optoelectronic components. The determination of the failure physical laws by analyzing degradation kinetics represents an additional step that can be integrated with the methodology proposed in this book.

Bibliography

[ADA 02] ADAMSON P., "Lead-free packaging for discrete power semiconductors", *Proceedings of JEDEC Conference*, pp. 316–351, 2002.

[AGI 90] AGIUS B., FROMENT M., *Surfaces, Interfaces et Films Minces: Observation et Analyse*, Dunod, 1990.

[ALL 08] ALLEN S.C., STECKL A.J., "A nearly ideal phosphor-converted white light-emitting diode", *Applied Physics Letters*, vol. 92, p. 143309, 2008.

[ALL 10] ALLEN N.S., *Photochemistry and Photophysics of Polymer Materials*, John Wiley and Sons, 2010.

[ANA 05] ANANTAWARASKUL S., *Polymer Analysis, Polymer Theory*, Springer, 2005.

[BAR 11] BARDSELEY, *et al.*, Solid-State Lighting Research & Development: Manufacturing Roadmap, DOE (Department of Energy), 2011.

[BEC 08] BECHOU L. *et al.*, "Measurement of the thermal characteristics of packaged double-heterostructure light emitting diodes for space applications using spontaneous optical spectrum properties", *Optics & Laser Technology*, vol. 40, pp. 589–601, 2008.

[BEN 87] BENNINGHOVEN A. *et al.*, *Secondary Ion Mass Spectrometry: Basic Concepts, Instrumental Aspects, Applications and Trends*, John Wiley & Sons, 1987.

[BER 94] BERSHTEÏN V.A., EGOROV V.M., *Differential Scanning Calorimetry of Polymers: Physics, Chemistry, Analysis, Technology*, Ellis Horwood, 1994.

[BER 11] BERNABE S., Mécanismes de dégradation des LEDs liés au matériau, à l'architecture et à l'intégration des puces et étude de leur impact sur la fiabilité des systèmes, Thesis, CEA-LETI, Grenoble, 2011.

[BRE 99] BRENNAN K.F., *The Physics of Semiconductors: with Applications to Optoelectronic Devices*, Cambridge University Press, 1999.

[BRU 92] BRUNDLE C.R. *et al.*, *Encyclopedia of Materials Characterization: Surfaces, Interfaces, Thin Films*, Gulf Professional Publishing, 1992.

[BYE 07] BYEON K.J. *et al.*, "Fabrication of two-dimensional photonic crystal patterns on GaN-based light-emitting diodes using thermally curable monomer-based nanoimprint lithography", *Applied Physics Letters*, vol. 91, p. 091106, 2007.

[CAO 04] CAO X.A. *et al.*, "Growth and characterization of blue and near-ultraviolet light emitting diodes on bulk GaN", *Fourth International Conference on Solid State Lighting, SPIE*, pp. 48–53, 2004.

[CAS 75] CASEY H.C. *et al.*, "Concentration dependence of the absorption coefficient for n-and p-type GaAs between 1.3 and 1.6 eV", *Journal of Applied Physics*, vol. 46, pp. 250–257, 1975.

[CEA 16] CEA–LETI, available at: www-leti.cea.fr/fr, 2016.

[CHA 03] CHANG C.S. *et al.*, "InGaN/GaN light-emitting diodes with ITO p-contact layers prepared by RF sputtering", *Semiconductor Science and Technology*, vol. 18, pp. L21–L23, 2003.

[CHA 05] CHANG K.M. *et al.*, "Investigation of indium–tin-oxide ohmic contact to p-GaN and its application to high-brightness GaN-based light-emitting diodes", *Solid-State Electronics*, vol. 49, pp. 1381–1386, 2005.

[CHE 06] CHEN W.S. *et al.*, "Rapid thermal annealed InGaN/GaN flip-chip LEDs", *IEEE Transactions on Electron Devices*, vol. 53, pp. 32–37, 2006.

[CHE 08] CHEN J.J. *et al.*, "Enhanced output power of GaN-based LEDs With nano-patterned sapphire substrates", *IEEE Photonics Technology Letters*, vol. 20, pp. 1193–1195, 2008.

[CHO 99] CHO Y.H. *et al.*, "Carrier dynamics of abnormal temperature-dependent emission shift in MOCVD grown InGaN epilayers and InGaN/GaN quantum wells", *MRS Internet Journal of Nitride Semiconductor Research*, vol. 4S1, p. G2.4, 1999.

[CHO 06] CHOI Y.S., KIM S.J., "Sapphire substrate-transferred nitride-based light-emitting diode fabricated by sapphire wet etching technique", *Solid-State Electronics*, vol. 50, pp. 1522–1528, 2006.

[CHO 08] CHO H.K. *et al.*, "Laser liftoff GaN thin-film photonic crystal GaN-based light-emitting diodes", *IEEE Photonics Technology Letters*, vol. 20, pp. 2096–2098, 2008.

[CHU 13] CHUANG S.-H., PAN C.-T., SHEN K.-C. *et al.*, "Thin film GaN LEDs using a patterned oxide sacrificial layer by chemical lift-off process", *IEEE Photonics Technology Letters*, vol. 25, no. 24, pp. 2435–2438, 2013.

[COL 09] COLE R.B., *Electrospray and MALDI Mass Spectrometry: Fundamentals, Instrumentation, Practicalities, and Biological Applications*, John Wiley and Sons, 2009.

[COR 06] CORDIER Y., "Filière HEMT AlGaN/GaN sur silicium", *Journées Nationales Microélectronique et Optoélectronique*, 2006.

[CRE 09] CREE LED Lamp Reliability Test Standard, pp. 1–21, 2009.

[CRE 10] CREE, Optimizing PCB thermal performance for Cree XLamp LEDs, Technical Article CLD-AP37 REV 1, pp. 1–20, 2010.

[CRH 16] CRHEA–CNRS, available at: www.crhea.cnrs.fr, 2016.

[DAK 06] DAKIN J., BROWN R.G.W., *Handbook of Optoelectronics*, vol. 2, Taylor & Francis, 2006.

[DAP 07] DA POS O. *et al.*, Colour rendering of white LED light sources, CIE Technical Report no. 177, 2007.

[DAS 07] DASS C., *Fundamentals of Contemporary Mass Spectrometry*, Wiley-Interscience, 2007.

[DAV 06] DAVID A. *et al.*, "Photonic crystal laser lift-off GaN light-emitting diodes", *Applied Physics Letters*, vol. 88, p. 133514, 2006.

[DAV 14] DAVID A., HURNI C.A., ALDAZ R.I. *et al.*, "High light extraction efficiency in bulk-GaN based volumetric violet light-emitting diodes", *Applied Physics Letters*, vol. 105, no. 23, p. 231111, 2014.

[DE 07] DE HOFFMANN E., STROOBANT V., *Mass Spectrometry: Principles and Applications*, Wiley-Interscience, 2007.

[DEG 04] DEGROOT J.V. *et al.*, "Highly transparent silicone materials", *Proc. SPIE 5517, Linear and Nonlinear Optics of Organic Materials*, vol. 4, pp. 116–123, October 15, 2004.

[DEM 07] DEMILO C. *et al.*, "Thermally induced stresses resulting from coefficient of thermal expansion differentials between an LED sub-mount material and various mounting substrates", *Proceedings of SPIE*, vol. 6486, p. 64860N, 2007.

[DEN 06] DENIS A. *et al.*, "Gallium nitride bulk crystal growth processes: a review", *Materials Science and Engineering Review*, vol. 50, pp. 167–194, 2006.

[DES 02] DESHAYES Y., Diagnostic de défaillances de systèmes optoélectroniques émissif pour applications de télécommunication: Caratérisations électro-optiques et simulations thermomécaniques, PhD Thesis, University of Bordeaux I, Bordeaux, 2002.

[DES 04] DESHAYES Y. *et al.*, "Estimation of lifetime distributions on 1550 nm DFB laser diodes using Monte-Carlo statistic computations", *Proceedings of SPIE*, vol. 5465, pp. 103–115, 2004.

[DES 10] DESHAYES Y. *et al.*, "Stark effects model used to highlight selective activation of failure mechanisms in MQW InGaN/GaN light-emitting diodes", *IEEE Transactions on Device and Materials Reliability*, vol. 10, pp. 164–170, 2010.

[DIA 01] DIAS F.B. *et al.*, "Internal dynamics of poly(methylphenylsiloxane) chains as revealed by picosecond time resolved fluorescence", *Journal of Physical Chemistry A*, vol. 105, pp. 10286–10295, 2001.

[DIN 15] DING Q.A., LI K., KONG F. *et al.*, "Improving the vertical light extraction efficiency of GaN-based thin-film flip-chip LED with double embedded photonic crystals", *IEEE Journal of Quantum Electronics*, vol. 51, no. 2, p. 6841005, 2015.

[DMI 00] DMITRIEV V.A., CHOW T.P., DEN BAARS, S.P. *et al.*, "High-temperature electronics in Europe", *TTEC Panel Report, International Technology Research Institute*, Chapter 2, pp. 1–197, available at: http://www.itri-us.com/ttec/hte_e/report/hte-europe.pdf, 2000.

[DUB 09] DUBOZ J.Y, RIBOT H., FEUILLET G., "L'éclairage à l'état solide au Japon", Mission Report, pp 1–39, available at: http://www1.eere.energy.gov/buildings/publications/pdfs/ssl/energystar_sslcriteria.pdf, 2009.

[DUF 06] DUFOUR P. *et al.*, "Microscopie à deux photons pour l'imagerie cellulaire fonctionnelle: avantages et enjeux", *Medecine/Sciences*, vol. 22, pp. 837–844, 2006.

[EKM 09] EKMAN R., *Mass Spectrometry: Instrumentation, Interpretation, and Applications*, John Wiley and Sons, 2009.

[ENE 08] ENERGY STAR, Program Requirements for Solid State Lighting Luminaires, Version 1.1, 2008.

[ERN 90] ERNST R.R. *et al.*, *Principles of Nuclear Magnetic Resonance in One and Two Dimensions*, Oxford University Press, 1990.

[FAN 07] FAN B. *et al.*, "Study of phosphor thermal-isolated packaging technologies for high-power white light-emitting diodes", *IEEE Photonics Technology Letters*, vol. 19, pp. 1121–1123, 2007.

[FRO 11a] FROST & SULLIVAN, World LED Lighting Markets, Research Report, 2011.

[FRO 11b] FROST & SULLIVAN, The LED Revolution and Key Opportunities for Lighting Companies in the Global Market, Research Report, 2011.

[FUK 91] FUKUDA M., *Reliability and Degradation of Semiconductor Laser and LEDs*, Artech House, London, 1991.

[GAL 06] GALAUP J.P., Spectroscopie Raman, 2006.

[GAN 06] GAN F., XU L., *Photonic Glasses*, World Scientific, 2006.

[GAO 08] GAO H. *et al.*, "Improvement of the performance of GaN-based LEDs grown on sapphire substrates patterned by wet and ICP etching", *Solid State Electronics*, vol. 52, pp. 962–967, 2008.

[GFI 09] GFIE, "Finitions des circuits imprimés: États des lieux, Avantages et inconvénients", *Les Cahiers de l'Industrie Électronique et Numérique*, pp. 8–11, 2009.

[GLO 07] GLOTCH T.D. *et al.*, "Attenuated total reflection as an in-situ infrared spectroscopic method for mineral identification", *Lunar and Planetary Science*, vol. XXXVIII, p. 1731, 2007.

[GLO 10] GLOBAL INDUSTRY ANALYSTS, Light Emitting Diode (LED): A Global Market Report, Reprot, p. 415, 2010.

[GOL 03] GOLDSTEIN J., *Scanning Electron Microscopy and X-Ray Microanalysis*, vol. 1, Springer, 2003.

[GRU 06] GRUNDMANN M., *The Physics of Semiconductors: An Introduction Including Devices and Nanophysics*, Springer, 2006.

[GU 04] GU Y., NARENDRAN N., "A non-contact method for determining junction temperature of phosphor-converted white Leds", *3rd Conference on Solid State Lighting*, pp. 107–114, 2004.

[GUI 87] GUILLET J., *Polymer Photophysics and Photochemistry: An Introduction to the Study of Photoprocesses in Macromolecules*, CUP Archive, 1987.

[HAM 00] HAMID S.H., *Handbook of Polymer Degradation*, Marcel Dekker, 2000.

[HAN 15] HAN J., LEE D., JIN B. *et al.*, "Optimizing *n*-type contact design and chip size for high-performance indium gallium nitride/gallium nitride-based thin-film vertical light-emitting diode", *Materials Science in Semiconductor Processing*, vol. 31, pp. 153–159, 2015.

[HEL 08] HELD G., *Introduction to Light Emitting Diode Technology and Applications*, CRC Press, 2008.

[HIR 96] HIRATA G.A. *et al.*, "High transmittance-low resistivity ZnO:Ga films by laser ablation", *Journal of Vacuum Science & Technology A*, vol. 14, pp. 791–794, 1996.

[HIR 01] HIRSCH L. *et al.*, "Nuclear microprobe analysis of GaN based light emitting diodes", *Physica Status Solidi A*, vol. 188, pp. 171–174, 2001.

[HIR 02] HIRSCH L. *et al.*, "Field distribution and collection efficiency in an AlGaN metal–semiconductor–metal detector", *Journal of Applied Physics*, vol. 91, p. 6095, 2002.

[HON 02] HONDA Y. *et al.*, "Growth of GaN free from cracks on a (111) Si substrate by selective metalorganic vapor-phase epitaxy", *Applied Physics Letters*, vol. 88, p. 222, 2002.

[HÖH 03] HÖHNE G. *et al.*, *Differential Scanning Calorimetry*, Springer, 2003.

[HON 04] HONG E., NARENDRAN N., "A method for projecting useful life of LED lighting systems", *Third conference on Solid State Lighting*, pp. 93–99, 2004.

[HON 08] HON S.J. *et al.*, "High power GaN LED chip with low thermal resistance", *Proceedings of SPIE*, vol. 6894, p. 689411, 2008.

[HOY 16] HOYA CORP. OPTICS DIVISION, *Optical Glass*, available at: http://www.hoyaoptics.com/pdf/OpticalGlass.pdf, pp. 1–22, 2016.

[HSU 10] HSU J., *LED Market Opportunities and Challenges*, IMS Research, 2010.

[HUA 09] HUANG H.W. *et al.*, "Investigation of GaN-based light emitting diodes with nano-hole patterned sapphire substrate (NHPSS) by nano-imprint lithography", *Materials Science and Engineering B*, vol. 164, pp. 76–79, 2009.

[HUY 05] HUYGHE S., Fiabilité des amplificateurs optiques à seminconducteur 1,55 μm pour des applications de télécommunication: Etude expérimentale et modélisation physique, PhD Thesis, University of Bordeaux I, 2005.

[IOF 01] IOFFE, *New Semiconductor Materials: Characteristics and Properties*, Ioffe, 2001.

[ISU 07] ISUPPLI CORP., "LEDs poised to drive a new lighting revolution", *LED Professional*, November 8, 2007.

[JAE 07] JAEGER R., GLERIA M., *Inorganic Polymers*, Nova Science Publishers, 2007.

[JOH 95] JOHANSSON S.A.E. *et al.*, *Particle-induced X-Ray Emission Spectrometry (PIXE)*, Wiley-Interscience, 1995.

[KIM 05a] KIM D.W. *et al.*, "A study of transparent contact to vertical GaN-based light-emitting diodes", *Journal of Applied Physics*, vol. 98, p. 053102, 2005.

[KIM 05b] KIM J.K. *et al.*, "Strongly enhanced phosphor efficiency in GaInN white light-emitting diodes using remote phosphor configuration and diffuse reflector cup", *Japanese Journal of Applied Physics*, vol. 44, pp. 649–651, 2005.

[KIM 05c] KIM S.J. *et al.*, "Vertical chip of GaN-based light-emitting diode formed on sapphire substrate", *Physica Status Solidi A*, vol. 202, pp. 2034–2039, 2005.

[KIM 08] KIM J.K. *et al.*, "Light-extraction enhancement of GaInN light-emitting diodes by graded-refractive-index indium tin oxide anti-reflection contact", *Advanced Materials*, vol. 20, pp. 801–804, 2008.

[KOI 92] KOIZUMIT H. *et al.*, "Light-induced electron emission from silicone oils", *Journal of Physics D: Applied Physics*, vol. 25, pp. 857–861, 1992.

[KRE 97] KREVELEN D.W.V., *Properties of Polymers*, 3rd ed., Elsevier, 1997.

[KUM 06] KUMAR R.N. *et al.*, "Ultraviolet radiation curable epoxy resin encapsulant for light emitting diodes", *Journal of Applied Polymer Science*, vol. 100, pp. 1048–1056, 2006.

[LAK 06] LAKOWICZ J.R., *Principles of Fluorescence Spectroscopy*, vol. 1, Springer, 2006.

[LAM 04] LAMBERT J.B., MAZZOLA E.P., *Nuclear Magnetic Resonance Spectroscopy: An Introduction to Principles, Applications, and Experimental Methods*, Pearson Education, 2004.

[LED 10] LEDMAN "LED reliability test standards", pp. 1–6, available at: http://optics.org/news/1/7/6, 2010.

[LED 10a] LEDMAN "LED market tops $10BN in 2010", available at: optics.org, 2010.

[LED 10b] LED INSIDE, *2009 Revenues of Global LED Package Vendors Reaches $8B*, LEDinside, 2010.

[LEE 98] LEE S.Y. *et al.*, "Deposition of low stress, high transmittance SiC as an X-ray mask membrane using ECR plasma CVD", *Journal of the Korean Physical Society*, vol. 33, pp. S156–S158, 1998.

[LEE 04] LEE B.K., GOH K.S., CHIN Y.L. *et al.*, Light emitting diode with gradient index layering, US Patent US6717362 B1, 2004.

[LEE 05] LEE K.J. *et al.*, "Growth and properties of blue/green InGaN/GaN MQWs on Si(111) substrates", *Journal of the Korean Physical Society*, vol. 47, pp. S512–S516, 2005.

[LEE 07] LEE T.X. *et al.*, "Light extraction analysis of GaN-based light-emitting diodes with surface texture and/or patterned substrate", *Optics Express*, vol. 15, pp. 6670–6676, 2007.

[LEE 09] LEE J. *et al.*, "GaN-based light-emitting diodes directly grown on sapphire substrate with holographically generated two-dimensional photonic crystal patterns", *Current Applied Physics*, vol. 9, pp. 633–635, 2009.

[LEI 91] LEI T. *et al.*, "Epitaxial growth of zinc blende and wurtzitic gallium nitride thin films on (001) silicon", *Applied Physics Letters*, vol. 59, p. 944, 1991.

[LES 98] LESTER S.D., MILLER J.N., ROITMAN D.B., High refractive index package material and a light emitting device encapsulated with such material, US Patent US5777433 A, 1998.

[LEW 01] LEWIS I.R., EDWARDS H.G.M., *Handbook of Raman Spectroscopy: from the Research Laboratory to the Process Line*, Marcel Dekker, 2001.

[LIF 99] LIFSHIN E., *X-Ray Characterization of Materials*, Wiley, 1999.

[LIN 10] LIN Y.H. *et al.*, "Development of high-performance optical silicone for the packaging of high-power LEDs", *IEEE Transactions on Components and Packaging Technologies*, vol. 33, pp. 761–766, 2010.

[LIU 09] LIU H. *et al.*, "Light extraction of GaN LEDs with 2-D photonic crystal structure", *Chinese Optics Letters*, vol. 7, pp. 918–920, 2009.

[LUO 05] LUO H. *et al.*, "Analysis of high-power packages for phosphor-based white-light-emitting diodes", *Applied Physics Letters*, vol. 86, p. 243505, 2005.

[LYM 90] LYMAN C.E., *Scanning Electron Microscopy, X-Ray Microanalysis, and Analytical Electron Microscopy: A Laboratory Workbook*, Springer, 1990.

[MAC 91] MACANITA A.L. *et al.*, "Decay of poly(phenylsiloxane) fluorescence emission. Kinetic parameters and rotational motion", *Macromolecules*, vol. 24, pp. 1293–1298, 1991.

[MAR 99] MARGALITH T. *et al.*, "Indium tin oxide contacts to gallium nitride optoelectronic devices", *Applied Physics Letters*, vol. 74, p. 3930, 1999.

[MAR 11] MARTINSONS C., Overview of the CITALED Program, Report, EU Commission, Brussels, 2011.

[MEN 06a] MENDIZABAL L., Fiabilité de diodes laser DFB 1,55 µm pour des applications de télécommunication: Approche statistique et interaction composant-système, PhD Thesis, University of Bordeaux 1, Bordeaux, 2006.

[MEN 06b] MENEGHINI M. *et al.*, "High-temperature degradation of GaN LEDs related to passivation", *IEEE Transaction on Electron Devices*, vol. 53, pp. 2981–2987, 2006.

[MEN 07] MENEGHINI M. *et al.*, "High temperature electro-optical degradation of InGaN/GaN HBLEDs", *Microelectronics Reliability*, vol. 47, pp. 1625–1629, 2007.

[MIN] MING T.C., *Reliability of Solid State Lighting with Case Study on Humidity Effect*, Nanyang Technological University.

[MIR 98] MIRABELLA F.M., *Modern Techniques in Applied Molecular Spectroscopy*, Wiley-IEEE, 1998.

[MIZ 86] MIZUTA M. *et al.*, "Low temperature growth of GaN and AlN on GaAs utilizing metalorganics and hydrazine", *Japanese Journal of Applied Physics*, vol. 25, pp. L945–L948, 1986.

[MOH 04] MOHAN J., *Organic Spectroscopy Principles and Applications*, 2nd ed., Taylor and Francis, 2004.

[MOR 08a] MORKOÇ H., *Handbook of Nitride Semiconductors and Devices: Materials Properties, Physics and Growth*, vol. 1, Wiley, 2008.

[MOR 08b] MORKOÇ H., *Handbook of Nitride Semiconductors and Devices: Electronic and Optical Processes in Nitrides*, vol. 2, Wiley, 2008.

[MOR 08c] MORKOÇ H., *Handbook of Nitride Semiconductors and Devices: GaN-based Optical and Electronic Devices*, vol. 3, Wiley, 2008.

[NAN 97] NANNY M.A. *et al.*, *Nuclear Magnetic Resonance Spectroscopy in Environmental Chemistry*, Oxford University Press, 1997.

[NAR 05] NARENDRAN N., "Improved performance White LED", *Proceedings of SPIE*, vol. 5941, p. 594108, 2005.

[NEW 86] NEWBURY D.E., *Advanced Scanning Electron Microscopy and X-ray Microanalysis*, Springer, 1986.

[NG 99] NG H.M. *et al.*, "Distributed Bragg reflectors based on AlN/GaN multilayers", *Applied Physics Letters*, vol. 74, pp. 1036–1038, 1999.

[NG 08] NG W.N. *et al.*, "Photonic crystal light-emitting diodes fabricated by microsphere lithography", *Nanotechnology*, vol. 19, p. 255302, 2008.

[NG 11] NG S., *Bright Prospects for LED Lighting*, Green Puchasing Asia, 2011.

[OH 08] OH T.S. *et al.*, "GaN-based light-emitting diodes on micro-lens patterned sapphire substrate", *Japanese Journal of Applied Physics*, vol. 47, pp. 5333–5336, 2008.

[OH 10] OH J.R. *et al.*, "Lowering color temperature of $Y_3Al_5O_{12}$:Ce^{3+} white light emitting diodes using reddish light-recycling filter", *Electrochemical Solid-State Letters*, vol. 13, pp. J5–J7, 2010.

[OID 10] OIDA, *Solid-State Lighting (SSL)-LED Roadmap Recommendations*, OIDA, 2010.

[PAD 66] PADOVANI F.A., STRATTON R., "Field and thermionic-field emission in Schottky barriers", *Solid State Electronics*, vol. 9, p. 695, 1966.

[PAI 89] PAISLEY M.J. *et al.*, "Growth of cubic phase gallium nitride by modified molecular beam epitaxy", *Journal of Vacuum Science & Technology A*, vol. 7, p. 701, 1989.

[PAN 09] PANTHA B.N. *et al.*, "Electrical and optical properties of p-type InGaN", *Applied Physics Letters*, vol. 95, p. 261904, 2009.

[PAS 10] PASKOVA T. *et al.*, "GaN substrates for III-nitride devices", *Proceedings of the IEEE*, vol. 98, pp. 1324–1338, 2010.

[PER 92] PERLIN P. *et al.*, "Raman scattering and x-ray-absorption spectroscopy in gallium nitride under high pressure", *Physical Review B: Condensed Matter and Materials Physics*, vol. 45, pp. 83–89, 1992.

[PER 05] PERKINELMER, FT-IR Spectroscopy: Attenuated Total Reflectance (ATR), Research Report, 2005.

[PIM 11] PIMPABUTE N. *et al.*, "Determination of optical constants and thickness of amorphous GaP thin film", *Optica Applicata*, vol. XLI, pp. 257–268, 2011.

[PIP 02] PIPREK J., NAKAMURA S., "Physics of high-power InGaN–GaN lasers", *IEE Proceedings Optoelectronics*, pp. 145–151, 2002.

[PIP 05] PIPREK J., *Optoelectronic Devices: Advanced Simulation and Analysis*, Springer, 2005.

[POW 93] POWELL R.C. *et al.*, "Heteroepitaxial wurtzite and zinc-blende structure GaN grown by reactive ion molecular beam epitaxy: growth kinetics, microstructure, and properties", *Journal of Applied Physics*, vol. 73, p. 189, 1993.

[RAB 95] RABEK J.F., *Polymer Photodegradation: Mechanisms and Experimental Methods*, Springer, 1995.

[RAB 96] RABEK J.F., *Photodegradation of Polymers: Physical Characteristics and Applications*, Springer, 1996.

[RAY 83] RAY S. *et al.*, "Properties of tin doped indium oxide thin films prepared by magnetron sputtering", *Journal of Applied Physics*, vol. 54, p. 3497, 1983.

[REE 06] REEH U., HOHN K., STATH N. *et al.*, Light-radiating semiconductor component with a luminescence conversion element, US Patent 20050161694 A1, 2006.

[REI 98] REIMER L., *Scanning Electron Microscopy: Physics of Image Formation and Microanalysis*, Springer, 1998.

[RES 05] RESHCHIKOVA M.A., MORKOÇ H., "Luminescence properties of defects in GaN", *Journal of Applied Physics*, vol. 97, p. 061301, 2005.

[RES 11] RESEARCH AND MARKETS, *OLED Lighting in Asia – 2011 – Market Opportunities and Developments*, Research and Markets, 2011.

[RIC 10] RICHMAN E., "LED reliability and current standards for measurement", *DOE SSL Workshop*, 2010.

[RIE 04] RIEGLER B. *et al.*, "Optical silicones for use in harsh operating environments", *Proceedings of Optics East*, pp. 1–15, 2004.

[ROS 02] ROSENCHER E., VINTER B., *Optoélectronique*, 2nd ed., Dunod, Paris, 2002.

[RUB 11] "RUBICON produces first 12-Inch sapphire wafers for LED manufacturing", *Semiconductor Today*, available at: http://www.semiconductor-today.com/news_items/2011/JAN/RUBICON_250111.htm, 25 January 2011.

[SCH 06] SCHUBERT E.F., *Light-Emitting Diodes*, 2nd ed., Cambridge University Press, 2006.

[SHA 03] SHAPIRO N.A. *et al.*, "Luminescence energy and carrier lifetime in InGaN–GaN quantum wells as a function of applied biaxial strain", *Journal of Applied Physics*, vol. 94, pp. 4520–4529, 2003.

[SHE 00] SHEU J.K. *et al.*, "Luminescence of an InGaN/GaN multiple quantum well light-emitting diode", *Solid-State Electronics*, vol. 44, pp. 1055–1058, 2000.

[SHI 06] SHIN M.W., "Thermal design of high-power LED package and system", *Proceedings of SPIE*, vol. 6355, p. 635509, 2006.

[SHI 15] SHI D., FENG S., ZHANG Y. *et al.*, "Thermal investigation of LED array with multiple packages based on the superposition method", *Microelectronics Journal*, vol. 46, pp. 632–636, 2015.

[SIL 07] "SILICONE Material Solutions for LED Packages and Assemblies", *Momentive Performance Materials*, pp. 1–9, 2007.

[SIN 01] SINGLETON J., *Band Theory and Electronic Properties of Solids*, Oxford University Press, 2001.

[SMI 05] SMITH E., DENT G., *Modern Raman Spectroscopy: A Practical Approach*, John Wiley and Sons, 2005.

[SMI 10] SMITH M.J., LED Market Segments, 2010.

[SPE 11] "Specifications for the Lighting Design Lab LED product lists", *LED product list developed for Puget Sound Energy, Seattle City Light, Snohomish PUD, Tacoma Power, the Energy Trust of Oregon, Bonneville Power Administration and Idaho Power*, 2011.

[STR 62] STRATTON R., "Theory of field emission from semiconductors", *Physical Review Letters*, vol. 125, pp. 67–82, 1962.

[SU 04] SU Y.K. *et al.*, "InGaN/GaN blue light-emitting diodes with self-assembled quantum dots", *Semiconductor Science and Technology*, vol. 19, pp. 389–392, 2004.

[SWA 95] SWAGER T.M. *et al.*, "Fluorescence studies of poly(p-phenyleneethyny1ene)s: the effect of anthracene substitution", *Journal of Physical Chemistry*, vol. 99, pp. 4886–4893, 1995.

[SZE 07] SZE S.M., NG K.K., *Physics of Semiconductor Devices*, 3rd ed., Wiley, 2007.

[TAR 08] TARTARIN J.G., La technologie GaN et ses applications pour l'électronique robuste, haute fréquence et de puissance, hal-00341009, pp. 1–16, 2008.

[THO 02] THOMAS N.L. *et al.*, "Widely tunable light-emitting diodes by Stark effect in forward bias", *Applied Physics Letters*, vol. 81, pp. 1582–1584, 2002.

[TOH] TOHNO M. *et al.*, "GaN-LED's on nano-etched sapphire substrate by metal-organic chemical vapor deposition", *Scivax Proceeding*.

[TOM 91] TOMIKI T. *et al.*, "Ce^{3+} centres in $Y_3Al_5O_{12}$ (YAG) single crystals", *Journal of the Physical Society of Japan*, vol. 60, pp. 2437–2445, 1991.

[TRA 99] TRAN C.A. *et al.*, "Growth of InGaN/GaN multiple-quantum-well blue light-emitting diodes on silicon by metalorganic vapor phase epitaxy", *Applied Physics Letters*, vol. 75, p. 1494, 1999.

[TRE 07] TREVISANELLO L.R. *et al.*, "Thermal stability analysis of high brightness LED during high temperature and electrical aging", *Proceedings of SPIE*, vol. 6669, p. 666913, 2007.

[TSU 04] TSUJI K. *et al.*, *X-Ray Spectrometry: Recent Technological Advances*, John Wiley and Sons, 2004.

[UED 96] UEDA O., *Reliability and Degradation of III-V Optical Devices*, Artech House, 1996.

[UEN 94] UENO M. *et al.*, "Stability of the wurzite-type structure under high pressure: GaN and InN", *Physical Review B: Condensed Matter and Materials Physics*, vol. 49, pp. 14–21, 1994.

[UN 09] UN, La population mondiale devrait dépasser les 9 milliards de personnes en 2050, UN News Centre, 2009.

[VAN 97] VAN DE WALLE C.G., "Interaction of hydrogen with native defects in GaN", *Physical Review B: Condensed Matter and Materials Physics*, vol. 56, pp. R10020–R10023, 1997.

[VAN 06] VANLATHEM E. *et al.*, "Novel silicone materials for LED packaging and opto-electronics devices", *Organic Optoelectronics and Photonics II*, pp. 619202.1–619202.8, 2006.

[VEY 10] VEYRIÉ D. *et al.*, "New methodology for the assessment of the thermal resistance of laser diodes and light emitting diodes", *Microelectronics Reliability*, vol. 50, pp. 456–461, 2010.

[VIC 89] VICKERMAN J.C. *et al.*, *Secondary Ion Mass Spectrometry: Principles and Applications*, Clarendon Press, 1989.

[WAD 94] WADA O., *Optoelectronic Integration: Physics, Technology, and Applications*, Springer, 1994.

[WAL 90] WALLS J.M., *Methods of Surface Analysis: Techniques and Applications*, CUP Archive, 1990.

[WAN 08] WANG H. *et al.*, "Active packaging method for light-emitting diode lamps with photosensitive epoxy resins", *IEEE Photonics Technology Letters*, vol. 20, pp. 87–89, 2008.

[WAR 90] WARREN B.E., *X-Ray Diffraction*, Courier Dover Publications, 1990.

[WAR 03] WARTEWIG S., *IR and Raman Spectroscopy: Fundamental Processing*, Wiley-VCH, 2003.

[WIL 89] WILSON R.G. *et al.*, *Secondary Ion Mass Spectrometry: A Practical Handbook for Depth Profiling and Bulk Impurity Analysis*, Wiley, 1989.

[WON 89] WONG C.P. *et al.*, "Understanding the use of silicone gels for nonhermetic plastic packaging", *IEEE Transactions on Components, Hybrids, and Manufacturing Technology*, vol. 12, pp. 421–425, 1989.

[WU 12] WU H.-H., LIN K.-H., LIN S.-T., "A study on the heat dissipation of high power multi-chip COB LEDs", *Microelectronics Journal*, vol. 43, pp. 280–287, 2012.

[WUU 06] WUU D.S. *et al.*, "Fabrication of pyramidal patterned sapphire substrates for high-efficiency InGaN-based light emitting diodes", *Journal of the Electrochemical Society*, vol. 153, pp. G765–G770, 2006.

[XIA 93] XIA Q. *et al.*, "Pressure-induced rocksalt phase of aluminum nitride: a metastable structure at ambient condition", *Journal of Applied Physics*, vol. 73, p. 8198, 1993.

[YAN 12] YANG C.T., LIU W.C., LI C.Y., "Measurement of thermal resistance of first-level Cu substrate used in high-power multi-chips LED package", *Microelectronics Reliability*, vol. 52, pp. 855–860, 2012.

[YUA 07] YUAN G.C. *et al.*, "Study of transmittance of ZnO film deposited on different substrate", *Guang pu xue yu guang pu fen xi Guang pu*, vol. 27, pp. 1263–1266, 2007.

[ZHA 08] ZHANG Y. *et al.*, "Temperature effects on photoluminescence of YAG:Ce^{3+} phosphor and performance in white light-emitting diodes", *Journal of the Rare Earths*, vol. 26, pp. 446–449, 2008.

[ZHE 14] ZHEN A., MA P., ZHANG Y. *et al.*, "Embeded photonic crystal at the interface of p-GaN and Ag reflector to improve light extraction of GaN-based flip-chip light-emitting diode", *Applied Physics Letters*, vol. 105, no. 25, p. 251103, 2014.

[ZHO 05] ZHOU S.Q. *et al.*, "Comparison of the properties of GaN grown on complex Si-based structures", *Applied Physics Letters*, vol. 86, p. 081912, 2005.

[ZOR 01] ZORODDU A. *et al.*, "First-principles prediction of structure, energetics, formation enthalpy, elastic constants, polarization, and piezoelectric constants of AlN, GaN, and InN: comparison of local and gradient-corrected density-functional theory", *Physical Review B: Condensed Matter and Materials Physics*, vol. 64, p. 045208, 2001.

Index

Printed in the United States
By Bookmasters